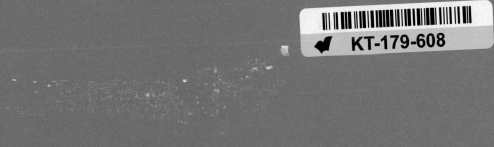

David Attenborough's

# first life

## A JOURNEY BACK IN TIME
## BY MATT KAPLAN WITH JOSH YOUNG

### INTRODUCTION BY
### DAVID ATTENBOROUGH

With material and images from the television series

Collins

# contents

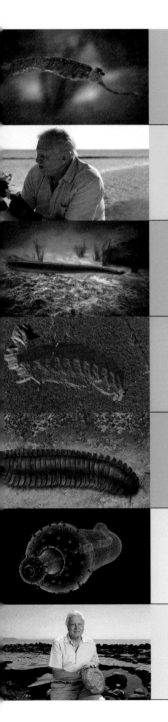

# first life
## INTRODUCTION

CHARNWOOD FOREST IS not a dense stand of trees. It is a forest only in the medieval sense of the word, that is to say, a wild relatively undeveloped patch of land where the soil is not rich or deep enough to make it worthwhile ploughing in order to grow crops. To me, as a boy, growing up in the nearby city of Leicester, it was the less interesting side of the county. My obsession was collecting fossils. In the eastern half of Leicestershire there were honey-coloured iron-rich limestone's which were full of fossils – bullet-shaped belemnites, wrinkled bi-valved shells the size and approximate shape of hazelnuts – and most beautiful of all – shells coiled like rams' horns, some no bigger than my fingernail, some – if I was lucky – six inches across. I collected all of these with great enthusiasm.

The rocks of Charnwood, however, I scarcely looked at. They were so old that they contained no fossils of any kind. Indeed, it was their very lack of fossils that defined them. The oldest fossiliferous rocks known to the pioneer geologists of Victorian times who established the basic outlines of their science were found in Wales and were accordingly given the name Cambrian. Charnwood's rocks, however, were even older. They were therefore called Precambrian.

Such fossil-free deposits, underlying the fossiliferous Cambrian, were soon recognised in many other parts of the world. Some of these were sediments that had been so compressed and distorted that no visible signs of living organisms could have survived. Some were crystalline igneous rock, solidified magma that had welled up from deep in the Earth's crust and could never have contained life of any kind. Charnwood's rocks, however, were of neither kind. They were layered, stratified and still relatively unchanged so that they might easily have contained fossils. But no one had ever found any. And I didn't even look.

Why the Precambrian lacked fossils was a great puzzle to geologists. In some parts of the world, there were Cambrian rocks lying immediately above them. The fossils they contained were varied and, for the most part strikingly different from any animals alive today. The simplest of them resembled fret-saw blades and were called graptolites. They acquired their name, which comes from a Greek word meaning writing, from a fancied resemblance of their little jagged lines to some kind of obscure primitive scribble. Their true nature was not established until the 1950s when delicate techniques of extracting their crushed remains from the surrounding matrix and examination of them through the microscope revealed that each tooth of the fret-saw blade was a socket that once held a tiny organism like a coral polyp.

But alongside these were very different creatures, trilobites. They looked like the woodlice or penny-sows that I knew from the garden. They had an armoured head and end section and in between a number of segments, each of which carried beneath it a pair of legs. They were, in short, quite complex creatures. To those Victorians who were trying to reconcile the discoveries made by geologists with the story of creation, as recorded in the Book of Genesis, it seemed that the first animals to appear on Earth were extraordinarily complicated.

Darwin and other scientists at the time did not accept, of course, that this could be so. Trilobites must have been preceded in the primeval seas by much simpler organisms. But even so, why had those hypothetical creatures left no trace? Perhaps there had been some world catastrophe that had eliminated all remains of previous animals from rocks everywhere. Perhaps the chemistry of the early oceans was such that its animal inhabitants were unable to extract the substances from it that were necessary for the secretion of shells. It was, in Darwin's words, 'a great mystery'.

*Charnia masoni:* the original fossil found by Roger Mason showing early life in an era where there was long believed to be none.

And then, in April 1957, an eleven-year-old boy named Roger Mason from the same Leicester grammar school that I had attended less than twenty years earlier, found a fossil in the Charnwood rocks. He and two friends, Richard Allen and Richard Blatchford, were climbing up the rocky sides of a disused quarry, close to a golf course at a place called Woodhouse Eaves. One of them noticed a strange leaf-like mark on the rock face, nearly eleven inches long. Curious marks had, it is true, been noticed on the Charnwood rocks before, but they had been dismissed as having been created by purely mechanical processes, by eddies and swirls that had left marks on the muds and sands that covered the ancient sea floor. But this was quite different. It was clearly a complicated structure with what appeared to be a central stem and branches coming from it on either side. It was placed in the middle of a very large slab of rock and it was simply not possible for the boys to chip it out and remove it. So Roger Mason, who already had a real interest in fossils, decided to make a rubbing of it. This he showed to his father who was a minister and taught part-time at Vaughan College. He in turn took it to Trevor Ford, who was in the geology department of Leicester University. He had doubts about the claim that it was the remains of a living thing but nonetheless went out to the quarry and clambered up to look at it. One glance was enough. 'My God, it is!' he exclaimed. He arranged for quarrymen to carefully cut out the fossil so that he could examine it properly in his laboratory. Superficially, it looked like the sea-pens that are found today on coral reefs, but Ford eventually decided that it was some kind of algal frond and published a scientific paper describing it in this way. And he gave it a name, *Charnia masoni*, commemorating both the place where it had been found and the schoolboy who had discovered it.

At the time of this discovery, science had not yet found a way of establishing the absolute age, in terms of millions of years, of any particular rock. Relative ages, however, could be deduced comparatively easily, providing the rocks concerned were sedimentary. Such rocks are the compressed and compacted sediments that originally accumulated at the bottom of ancient seas. Clearly, if two layers of such sediments lie one above the other (and are relatively undisturbed) then the lower layer must be older than the upper. Such sediments may contain the remains of marine creatures, such as shells, corals and the bones of fish. Most species change over time, evolving into new forms. So even if such sediments have become folded and crumpled as the drifting continents collide with one another, a particular layer can be traced from one exposure to another by recognising the particular species of fossils that it contains.

Using this method, the founding fathers of geology were able to work out the complete sequence of the sedimentary rocks that form the lands of Britain, and they had established that the rocks of Charnwood

*Charnia* was a marine organism that lived at the bottom of the ocean in the Precambrian period.

Forest were without question Precambrian. The discovery of an incontrovertible fossil in them, therefore, had important implications worldwide. Ten years earlier, very similar fossils had been discovered by a geologist working in the Ediacara Hills of southern Australia. They were known to be very ancient, but the current assumption that Precambrian rocks were devoid of fossils of any kind, meant that they were classified as Cambrian. The discovery of an incontrovertible fossil in the Precambrian rocks of the Charnwood Forest destroyed any such assumption. So the Ediacaran rocks were reclassified as Precambrian. The palaeontological world took a deep breath and re-adjusted its ideas. Old beliefs and assumptions were re-examined. New searches were made in ancient localities and geologists everywhere recognised that a whole new era in the history of life was ready for rediscovery.

In the fifty years that have passed since that discovery in Charnwood, *Charnia* has been recognised in rocks as far apart as Newfoundland, Siberia and the White Sea. Living organisms have been found that explain strange swirling concentric shapes that came from even earlier rocks than those in Charnwood and a whole chronology of Planet Earth dating back to over two billion years has been established. This book is about those extraordinary discoveries, how they were made and the people who made them.

# A Boyhood Passion

" Everyone has a passion, a favourite hobby, and mine has always been collecting fossils. My love for fossils developed as a boy, growing up in Leicester. There are Jurassic limestone outcrops in the eastern half of Leicestershire, and I used to cycle out to them after school and at weekends to search for fossils. The limestone that contained these fossils is particularly rich in iron and, at that time, it was still used as iron ore, so many of the nearby quarries were still in action. There were also other quarries that were overgrown and disused; they were my treasure field. As the rock layers weathered and eroded, the fossils within were gradually exposed.

Finding my first fossil was one of the key moments of my life. I can describe the moment in detail because I've relived the experiences of discovering fossils many times since. The thrill of discovery has never worn off. I remember vividly the moment when I first hit a lump of limestone with a hammer, splitting it apart. There, perfect in every detail and glinting as if it had just been polished, was a coiled seashell, 8 to 10 cm (3 to 4 in) across. It was an object of breathtaking beauty, and I was the first to see it since its occupant had died 200 million years earlier. It was an ammonite, an extinct creature related to a modern-day nautilus, which sailed to great depths through the Jurassic seas. The limestone in which it lay was made up of the solidified, compressed mud which had accumulated at the bottom of those seas.

Finding fossils was just the first part of the adventure. Next was cleaning and identifying the fossils. I used to take my fossils down to the New Walk Museum in Leicester where a very kind geologist called H. H. Gregory would help me to identify them. In return, I would help out at the museum in my school holidays, classifying and sorting the museum fossil collection. Throughout this time I learnt a great deal about fossils and developed my fascination with natural history, a fascination that has never left me."

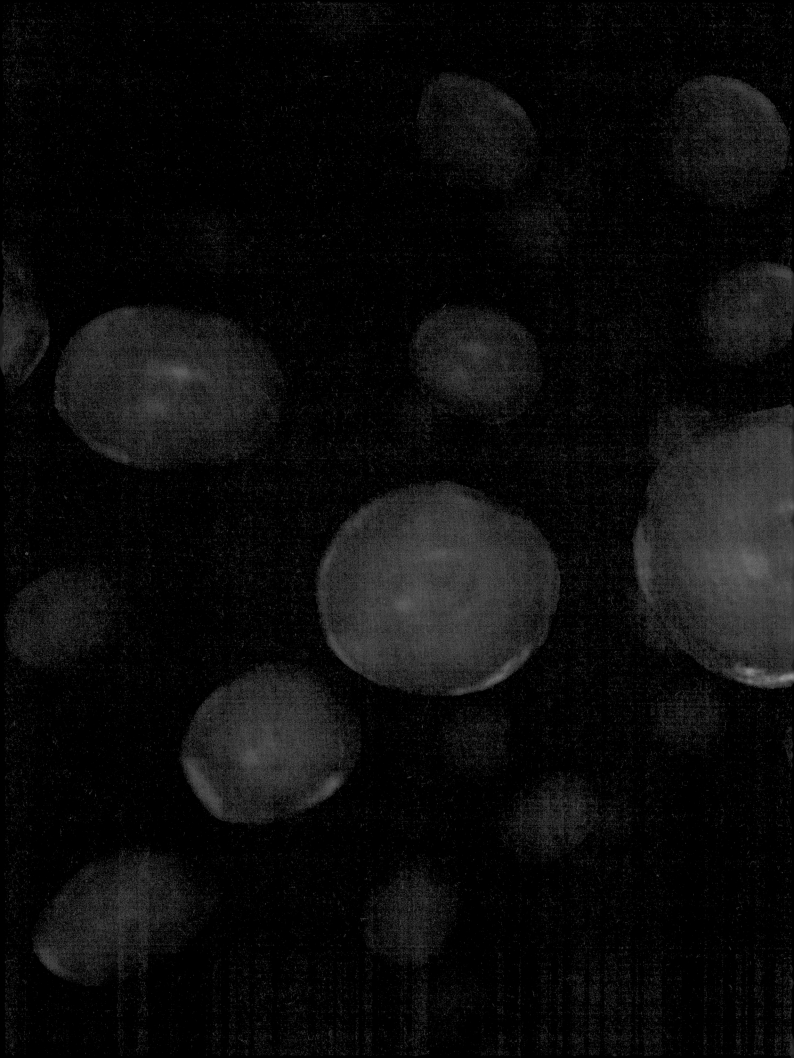

CHAPTER 1 · SOMETHING FROM NOTHING

THE HISTORY OF life is the most spectacular epic tale. Its storyline spans billions of years, from the dawn of life in Earth's ancient and hostile environment to the invasion of land by the first terrestrial organisms. The journey of life is full of bizarre, primitive creatures, from giant, bracken-like fronds to five-eyed predators with corkscrew mouths. Catastrophes as well as happy accidents pepper the rich evolutionary journey of animals. There are lineages that didn't make it, just as there are evolutionary innovations that persist today in the blueprints of us all. Without these twists to the tale, life as we know it simply wouldn't exist. This is perhaps what is most incredible about the life that we see all around us today – the fact that it very nearly didn't happen.

We now know that all life on Earth today evolved from a common ancestor that first appeared roughly 3.5 billion years ago, when the Earth was a very different place. The continents were only just beginning to form, the days were just four hours long because the Earth spun much more quickly, and the little land that existed above the waves was hostile, volcanic and lifeless.

It was in this alien world that life began deep under the oceans. Nobody knows exactly how or when it happened, but scientists believe that volcanic vents at the bottom of the sea may have supplied the ingredients needed to build the first cells. Some of these cells began to make food from sunlight, growing in eerie towers that stretched towards the sun. Tracing the evolutionary tree, we now see that these primitive organisms were the very earliest ancestors of plants.

Remarkably, for three-quarters of Earth's history, single-celled life was all there was. After such a speedy start, life on Earth operated on go-slow for the next 3 billion years – a time that period palaeontologists jokingly refer to as the 'boring billions'. Then, around 600 million years ago – the time in which *Charnia* lived – things changed dramatically.

The catalyst for change was the world's greatest-ever ice age, when the Earth was almost completely covered in ice. It was the thawing of the planet that saw the rise of myriad new and complex life forms. They burst onto the world scene in the geological equivalent of the blink of an eye, each trying out newly evolved ways of dealing with the problems of living.

This preliminary stage in the story of animals was soon superseded by the Big Bang of evolution, the famous Cambrian Explosion of 542 million years ago. This diversity created in the last half a billion years wouldn't have been achieved without the evolution of one particularly important feature of animal life: sex.

But, of course, lots of other things happened between prehistory and today. Life saw the evolution of the first hard body parts in the first predators, and the first eyes in the first prey. And, in some of the most important transformations of all, natural selection gave the world the first backbones and the first feet. There are surprises in animal evolution, too, ones that reveal the deep relationship and common ancestry of all life.

———

But the mysteries surrounding the origins of animal life begin long before the evolutionary origins of the first eyes. In fact, the first major hurdle that scientists had to overcome in order to piece together the evolutionary journey of animal life was to understand the increasing complexity of the first cells, the very building blocks of every living creature – humans, sea urchins, *Tyrannosaurus rex* or any other extinct or living organism. These first cells soon combined together to form a single larger organism that could take in food in order to provide each individual cell with the resources that they needed to survive. So where and when did such biological teamwork evolve? Why did it happen?

These and other fundamental questions about where the life we see around us comes from, and why it is the way it is, are incredibly difficult for palaeontologists to answer.

The times when the evolutionary events that were shaping the first cells took place were very, very long ago. We are grappling here with some extreme ages. It is difficult enough for people, who typically live for less than a century, to comprehend the concept of a thousand years. Yet it is possible to visualize. If every generation lived for an average of 50 years, a thousand years would have been 20 generations ago; a time when Viking ships prowled the coasts of Europe, food was scarce and slavery was acceptable. Foreign as it may seem, this was a world where a recognizable form of civilization existed.

Ten thousand years – or 200 generations – reaches back to the earliest days of human society. Writing had not yet been invented, currency did not exist and even the basic concepts of farming had not begun. This is about the end of the Stone Age, known as the Neolithic period.

We can still conjure up some kind of sense of what it may have been like to live in the Stone Age, but already we are losing our sense of time. So when we think of a million years ago (that would give a family tree with 20,000 generations), it is virtually impossible to comprehend. Although a colossal number, it is a mere blip in the history of the Earth.

A trilobite specimen discovered in Morocco. Trilobites were probably the most advanced form of life on the planet at the start of the Cambrian period.

Excavating trilobite fossils on Mount Issamour in Morocco has become a major industry. It can take weeks to prepare the specimens.

To appreciate the sense of scale involved in the history of our 4.5-billion-year-old planet, we should imagine it as being just one year old. On that timescale, humans appeared 25 minutes ago. Even *Charnia*, one of the oldest complex organisms we know, is a mere 7 weeks old. The Vikings were here just 8 seconds ago.

The time since the last days of the dinosaurs, 65 million years ago, is still only 5 days and 6 hours. In comparison, the first fish appeared some 470 million years ago, the equivalent of 38 days, having lived for roughly 10 per cent of the Earth's history. And the appearance of the first life on Earth? Scientists estimate this amazing event to have occurred somewhere between 3.8 billion and 3.5 billion years ago, so that's an extraordinary 296 days ago in our hypothetical year.

These extreme ages make finding evidence of ancient life difficult because our planet is a restless one. The surface of the planet is constantly in motion due to a phenomenon called plate tectonics, and it is constantly being eroded by wind and water. This combination of natural forces is highly destructive and, as a result, the evidence of most of the Earth's ancient history has been erased.

Finding and unearthing fossils is labour-intensive, but the whole process, from fossilization to discovery, relies on one thing: something hard. But what if an animal does not have bones or hard shells incorporated into its body? There are plenty of boneless animals out there today, like jellyfish and sea anemones, so how on earth does a palaeontologist go about studying their ancient relatives? It is certainly a much tougher job but not impossible. Soft-bodied animals can be fossilized but only under much more specific conditions.

First, they must fall to the bottom of the ocean or sea that they are in and then not be eaten by scavengers or carried away by currents. Next, they must be covered by sediment, but not any sediment will do. To be properly preserved, these animals must be covered with sediment that is particularly fine, like mud or volcanic ash. If such fine sediment is present, and if the environment is calm enough so that the sediment can gently settle, then the blanket-like layer of sediment that accumulates can record an impression of the dead, boneless organism.

Knowing that soft-bodied fossils are mainly found in mud and ash rock formations from ancient ocean floors, palaeontologists spurn sandstone, since the large- and coarse-grained sand would never form an impression of the animals they are studying. Instead, they focus their efforts on rare rocks formed near ancient volcanoes that spewed tonnes of ash into the sea or near areas where ancient rivers dumped their waters into the ocean.

Indeed, there are only a few locations like this on Earth, these mud- and ash-filled ocean floors that have since been forced up by the movement of the Earth's crust to create dry land. In terms of the fossils

they yield, they are seen as sites that hold great secrets and provide an amazing snapshot of life on this planet from inconceivably ancient times.

Even the astounding quality of preservation that you find at these sites is still not enough for palaeontologists, who talk in terms of hundreds of millions of years and who are searching for evidence of the simplest life forms.

We are surrounded by multicellular organisms, like plants, animals and fungi. But if you take in the sum total of all life on Earth, cellular teamwork is actually rather rare. The majority of living things, such as bacteria, algae, plankton and the like, are actually made up of just one microscopic cell that does everything on its own. We cannot see them with the naked eye because they are so small, but single-celled life is everywhere.

Quite astonishingly, some of these single-celled organisms have made it into the fossil record. Such fossils are extremely rare, but they can be formed when the cells are exposed to silica dissolved in water. Silica, which is the main ingredient in glass, is a common enough compound on the planet, and most of it is now found in sand or in the skeletons of animals like sponges.

In the past, dissolved silica was much more common in the sea. Occasionally, when the chemical conditions were just right, it could harden in the water. Solidified crystals of silica created a casing so that it effectively functioned as a time capsule and kept the organisms preserved as fossils for billions of years.

Have we gone back far enough? Do these ancient cells give us the picture of the first life on Earth? Unfortunately not. They may be the oldest of fossils, but they are far too fully formed and far too complex in their life patterns to have been the first life to appear on Earth. It is time for the palaeontologists to come in from wild expeditions and rock climbing adventures, put away their trowels and head for the chemistry laboratory.

Separating what is alive from what is not seems simple at first glance, but even this most basic of classifications has its complications. Scientists say that living things must have four special characteristics: they must eat (even if it is a plant making its own food), be able to reproduce themselves, respond to their environment (evolve), and demonstrate an organized body structure.

Based on these criteria, we can ask: is sugar alive? If you have ever grown sugar crystals on a string, you will know that crystals of varying shapes will readily grow and develop into beautiful, organized, geometric structures. These crystals undoubtedly have an organized body structure and they appear to reproduce and grow, but they will never show any sort of behaviour and they will never need to eat. You cannot starve a crystal. But deny even the smallest single-celled organism food and it will, eventually, die.

Armed with a basic understanding of what traits simple life must have had, scientists in the 1950s started combining chemicals in their labs that they thought were likely to have been present in the Earth's oceans and atmosphere more than 3.5 billion years ago. These initial experiments involved hydrogen, ammonia and methane, as well as electric sparks to simulate the lightning that would have shot down from the prehistoric skies.

 It is a very simple soup: three gases and a bit of added fizz-bang. But it was enough to power some extraordinary chemistry. Rather remarkably, these early experiments produced complex compounds called amino acids. When linked together, they form the critical proteins that make up all living cells.

Inspired, the scientists tried using different starting chemicals, and variations in exposure to the types of hot and cold temperatures that would have occurred on the Earth long ago. Out of these noxious concoctions, they yielded even more amino acid varieties, as well as many different sugars, phosphates and nucleic acid bases that are the building blocks of DNA, the crucial instruction for the structure and behaviour of cells. These experiments suggested that the components of life really could have come from the chemical odds and ends that were present on our ancient planet.

There is yet more evidence that the essential components of life can form spontaneously. Every now and then, these compounds quite literally fall out of the sky, carried on meteorites. Studies of meteorite chemistry have revealed that their surfaces contain a reasonable number of carbon-based compounds (called organic compounds by chemists). In fact, some meteorites have been found to carry more than 70 amino acids, of which six are known to be vital components of biological proteins. They also contain sugars and fats that are common in living cells and even some components of DNA. With these observations, many researchers suggest that if this kind of chemical manufacturing can happen on asteroids in the voids of space, it could also have taken place in the early Earth environment. The trouble is, just because these components were present does not explain why they all came together to form a cell in the first place.

––––––

Nobody is sure where all of the chemical reactions that created life actually took place 3.5 billion years ago. Most of the scientists working in this area are looking for life's birthplace in watery places. But there was an awful lot of water on the ancient Earth, as there is today, and the ancient oceans and lakes have been reshaped many times over the last few billion years. Where life actually found a foothold may, therefore, never be known, but scientists cannot help hypothesizing – especially when they discover species in our modern world that bear an uncanny resemblance to what they think the very first life forms could have looked like.

In the 1970s, marine biologists discovered something they had previously thought impossible. In the superheated waters around deep underwater volcanic vents, they observed entire communities of organisms living in the supposedly uninhabitable environment. These communities were living on organisms that could survive perfectly well without sunlight.

It was a eureka moment. Palaeontologists got incredibly excited when they heard about these cells that could live off the swirling clouds of minerals being spewed out of the vents. Dramatic theories also swirled around, about how life must have been formed deep on the ocean floor in these active hot spots, where the Earth was releasing both chemicals and heat into the water.

Studies of these hydrothermal vents, which sometimes look like smoking black chimneys in the sea floor, hence their name 'black smokers', show that the vents themselves are sterile. The water around them is searingly hot, over 300° C (572° F), filled with chemicals that would be toxic to most forms of life today and enshrouded in complete darkness. Yet this superheated water mixes with the surrounding cold sea water, and a short distance from the vent, where the water is a relatively mild 120° C (248° F), you find bacteria that can cope with the temperatures and thrive on the toxic chemicals, using them as a source of energy.

For around 3 billion years, there was only single-celled life on Earth. Multicellular life has only evolved in the last 600 million years.

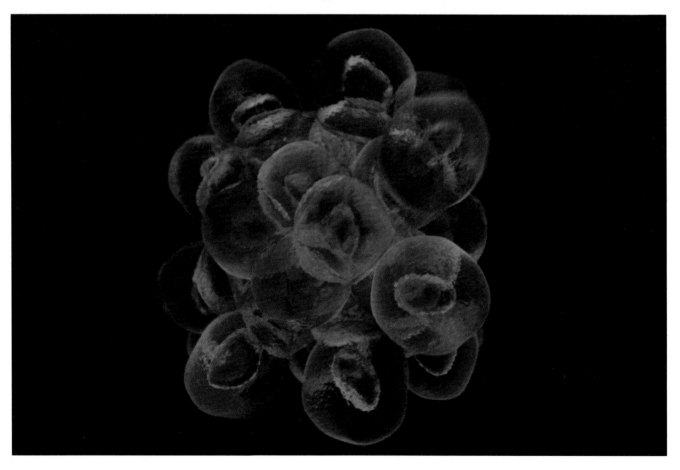

Have we finally found the place of the first life on Earth, the place where cells first formed, divided and gave birth to the second generation of life? It depends on the vent.

The hottest vents are created by magma that is burbling up from deep within the Earth. These black smokers burp out dark-coloured gases into the water, and scientists studying them today find that the cooler margins surrounding the vents are teeming with life. Lab work experimenting with the basic chemical compounds of early life under black smoker conditions shows that the nucleic acids that were likely to have been involved in the formation and subsequent replication of early cells would have been utterly destroyed. Nucleic acids, as well as the DNA that they form, just cannot tolerate such intense heat.

—

*It was a eureka moment. Palaeontologists got incredibly excited when they heard about these cells that could live off the swirling clouds of minerals being spewed out of the vents. Dramatic theories also swirled around, about how life must have been formed deep on the ocean floor in these active hot spots, where the Earth was releasing both chemicals and heat into the water.*

But there is a second type of vent, first discovered as recently as 2000, that is nowhere near as hot, and it is an intriguing alternative birthplace for that first life. These vents, or chemical seeps as they are more commonly known, form when certain kinds of rocks react with sea water. As the rock reacts, it cracks and lets in more water. This inrush of water then reacts further inside the crack, so the crack extends. Water penetrates deeper and deeper into the rock.

The scale of this process is absolutely astonishing. Scientists believe that the volume of water that has poured into rock and reacted in this way is equal to the volume of the oceans themselves.

Sometimes the water cuts cracks so deep that it becomes superheated from the molten magma inside the Earth. Indeed, this water boils and leads to the release of copious amounts of gases, such as hydrogen, methane, ammonia and hydrogen sulphide – all chemicals that would have been useful to the formation of early life. The hot water from the chemical seep rises, just like hot air rises to lift a hot-air balloon, and, in doing so, carries this mix of dissolved chemicals back to the surface, where it breaks through an underwater vent. By the time the hot-water mix gets back up to the sea bed, it has cooled down significantly. The vent water is warm, but not superheated, and rich in life-friendly chemical compounds.

Some of the minerals carried by the warm water tend to solidify as they mix with the cold ocean water and form tall complex structures riddled with tiny bubbles and compartments. At a distance, they look very much like spires. When researchers discovered the first of these vent sites located in the middle of the northern Atlantic Ocean, they named it Lost City.

But is there life in this Lost City? Today, there is a variety of animals, including snails, molluscs and worms, making their homes in these wonderful structures. But take time to explore the gleaming white spires, too, and peer in at the tiny compartments. They are perfectly sized for a snug-fitting cell. With so many rich chemicals coming up from the earth, the compartments could have been ideal places for chemical compounds to concentrate and combine and form early life in a relatively well-enclosed environment.

Scientists have analysed the cell-sized pores at locations like Lost City and found that these chambers concentrate the life-friendly compounds bubbling up from the vent, creating almost ideal reaction vessels for first life. Moreover, the chemicals that seep out of the vents are different from the chemicals in the surrounding sea water, and this chemical imbalance creates an electrical difference, or charge, a bit like that found inside a battery. This electrical potential could have provided energy to drive the chemical reactions taking place inside the spires of ancient versions of Lost City. With electrical power and a good concentration of vital ingredients, the self-sustaining chemistry of life may well have begun in the cosy compartments of such ornate cathedral spires.

The hydrothermal vent theories for the origins of life are fascinating and have been around since the 1980s, but they are not the only suggestions to explain life's humble beginnings. There are some researchers who think that early life probably evolved far from any hydrothermal vents and chemical seeps. They believe that life began on the land.

In 2001, the long-held view of oceans being the birthplaces of life fell into question with a series of experiments that tinkered with the role that salt might have played in the formation of the protective coverings around early cells.

Today, cells have envelopes called membranes surrounding them. They keep all the important bits of the cell inside and prevent all the nasty and harmful bits of the outside world from entering. If chemical seeps were the place where life formed, life, at some point, had to sally forth and move from its little chambers inside the mineral structures of places like Lost City and start living inside the types of membranes found around living cells today. There are ways to explain this transition, but some scientists argue that explaining the transition is

A white spire in the Lost City. With so many rich chemicals coming up from the Earth, the compartments could have been ideal places for chemical compounds to concentrate and combine and form early life in a relatively well-enclosed environment.

# Out of this World

Could the chemistry of life have occurred elsewhere in space? Certainly no living things can be found on Mars, Venus, Mercury or the moon. Venus and Mercury are too hot, while Mars and the moon are too cold because they are both far from the sun and so small that their own internal heat died out long ago. The giant gas planets Jupiter and Saturn, cold and inhospitable, also appear to be out of the question.

However, because Jupiter is so big (its mass is more than 2.5 times the mass of all the other planets in the solar system), it generates a lot of gravitational pull and has captured over 60 moons into its orbit. Some of these moons are quite big, and one, named Europa, has caught the eye of space scientists looking for signs of life.

Analysis by passing spacecraft over the years has revealed that Europa is covered in a smooth layer of ice. It has an iron core, as Earth does, and an atmosphere dominated by oxygen. With the extreme gravitational pull exerted on it by Jupiter, it is possible that Europa's iron core is being squeezed and that this, in turn, is generating heat. If there is internal heat being generated in the core, this heat is almost certainly moving out from the centre of the moon and travelling towards the surface, just as it does on Earth. Since Europa's surface is covered in ice, at the point where the internal heat meets the ice, there is likely to be a layer of water.

Much more exploration of Europa needs to be conducted to work through these theories. If they prove correct, we could one day find some aliens really close to home – organisms living in deep-sea vents in an ice-covered ocean.

Europa, one of Jupiter's many moons, which almost certainly possesses a vast ice-capped global ocean.

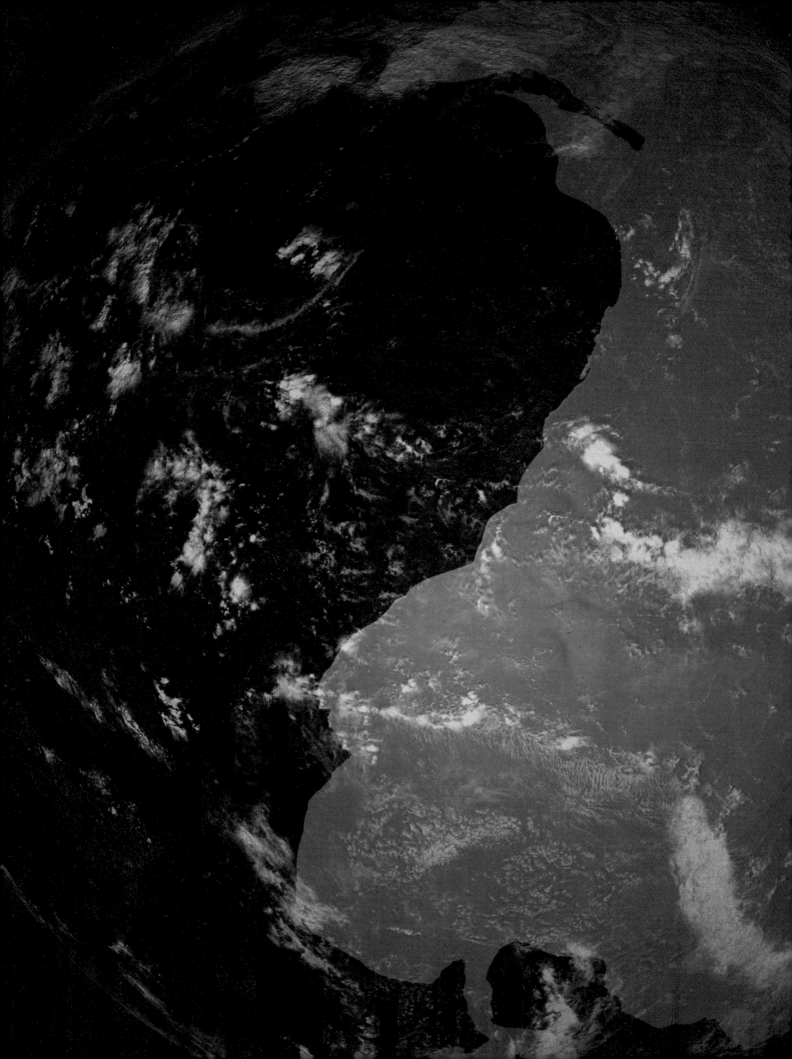

not necessary. Instead, these researchers suggest that the chemistry of life began encased in the oily membranes from the very start.

The fatty membranes of modern cells are complex things with clever systems that let them scan and identify chemicals and objects in the surrounding environment. If they identify a chemical they need, they let it into the interior of the cell, but if it doesn't fit, then the cell membrane stays tightly shut.

However unlikely it is that the very first cells started life with such a complex protective barrier, it is suggested that a primitive version of this type of membrane may have characterized the first living cells. While it is not agreed how they might have formed, some scientists argue that a single drop of an ancient oily substance stirred into water would have done the trick.

Fats and oils (or lipids as they are called by biochemists) are made of molecules that have dual personalities. They are composed of a head, which is attracted to water, and a tail, which repels it. So when certain types of lipids are dropped in water and mixed around, the molecules arrange themselves so that all of the heads face the water, and all of the tails are kept away. You end up with a lipid ball, with all the heads facing out and all the tails tucked safely in the interior of the ball.

The components that make up these oils have been detected on meteorites. Perhaps if one such lump of extraterrestrial rock came speeding through the atmosphere and splashed down into the sea, the extraterrestrial lipids would have formed these lipid membrane spheres in the ancient Earth's oceans. Scientists playing with these chemicals have shown that when the tiny spheres form in waters where life's critical components, such as proteins, phosphates, sugars and DNA are in high concentrations, the spheres can engulf them and, as a result, increase in size. This process of chemical capture concentrates the chemicals that get trapped inside, which encourages the chemicals to interact through a series of reactions. These little oily balls bobbing in the ocean would have effectively turned into drifting chemical factories.

With a little internal organization, these tiny chemical factories could have maximized the rate of reactions and absorbed materials from the outside environment as fuel to keep them going. This consumption of raw materials might have drawn the drifting factories to areas where the chemical fuels were present in high concentrations.

It sounds quite plausible until you try it out in the lab. Scientists have long assumed that life arose in the ocean because this is the main source of water on the planet. But when they tried to create these oily capsules in the lab, they discovered that they simply fell apart in salty water, even with salt in much lower concentrations than is present in the sea today. But in fresh water the spheres form easily.

# Scientific Debate

"We may think that scientists know all the answers, but that simply isn't the case. Different points of view and contrasting evidence often bring about scientific debates, debates which can continue back and forth for years, even decades. Most scientists are happy to disagree and argue their case in a convivial manner, but on some occasions, arguments over science can become pretty heated. Sometimes academics defend their views with such passion that huge rows erupt. There are cases of this happening throughout scientific history.

A great example is the row over Darwin's theory of natural selection as the cause of evolution. Darwin's theories are now the accepted view, but when he published his theory, it caused uproar amongst the scientific community. His most outspoken critic was Richard Owen, an eminent biologist and palaeontologist and a highly influential man. In a rather underhand attack, Owen wrote an anonymous article which described Darwin's theory as 'inconceivable', whilst praising his own theories. Darwin and Owen never really saw eye to eye and Darwin even wrote of Owen, 'I used to be ashamed of hating him so much, but now I will carefully cherish my hatred & contempt to the last days of my life.' Certainly strong words!

Debate about evolution continues to this day. There are many famous arguments about the nature of fossils, about their age and about what they represent. It's this grit, and the discovery of new fossils, that make science really exciting."

**On location in Australia during the filming of *First Life*.**

This was baffling because oceans long ago were probably much saltier than they are today because salt, over time, has accumulated on the continents and reduced the overall salt content in the oceans. While the exact salt concentrations of the ancient oceans are not known, the speculation is that they were between 1.5 and 2 times saltier than modern oceans.

So how could protective, oily capsules have ever formed around and protected the early components of life? The idea of creating life in hot, salty water at the bottom of the ocean seemed far less convincing. Maybe, then, life first formed in a freshwater environment. But where?

Inland lakes and rivers might have been a possible location, and there is no reason to discount them as places for life's formation, but freshwater lagoons on tropical volcanic islands probably hold the most promise. Ancient lagoon waters, shallow and heated by the sun, would have been warm, but not hot. The warmth would have accelerated the rate at which early cellular reactions took place and increased the chances of cells actually forming. The volcanism of the island would have been an important factor as well. Just as deep-sea vents belch out tonnes of chemicals that are useful to living things, volcanoes on the land can also release nutrients into the fresh water. Moreover, eruptions from volcanoes can create thick volcanic storm clouds that rain down water, ash and chemicals – and we know that those essential bolts of lightning gave early life's chemical reactions that vital kick.

The combination of warm temperatures, electricity, fresh water and a wide variety of life-friendly chemicals from deep within the Earth makes the tropical island idea enticing. But hard evidence to prove that such islands were the homes of first life remains elusive.

——

There is one final theory about the origins of the first life forms that is being considered. It comes from a handful of researchers who believe that there is sufficient evidence to suggest that life originally came to our planet from outer space.

As much as it might sound like science fiction, there are experiments that demonstrate how many of the critical life compounds can form in ice on dust grains in environments with no air present at all. Comets, like oversized snowballs flying around the sun, fit the bill perfectly as extraterrestrial factories for life chemicals. It is not impossible to imagine that small pieces of comet could break off and eventually land on Earth.

While the idea is indeed possible, few scientists believe that it's a plausible explanation for the origin of life. Dozens of experiments have all shown that the components for life could have easily arisen on Earth with no extraterrestrial input required. So, intriguing as the idea of comets seeding the earth might be, there are much simpler explanations that are firmly rooted to this planet alone.

We have theorized, speculated and contemplated all the reasonable explanations for how the first life on this planet arose.

We saw a schoolboy make an ordinary discovery that turned out to be an extraordinary scientific advancement. We have sunk deep into the ocean, skirted around superhot vents and chemical seeps, and we have explored volcanic islands and warm lagoons. Somewhere, in places like these, the chemistry that brought together the first forms of life began.

However and wherever it may have occurred, life on our planet first formed as little chemical factories in Earth's early waters. They possessed organized structures and the ability to feed, as well as behaviours that drew them to areas where they could absorb their vital chemical food. But what about reproduction? How did the very first of these chance organizations of molecules and chemical compounds manage to replicate?

*Scientists love big questions, and they do not get much bigger than those about the earliest days of life on Earth. Even when a living cell that is organized, able to absorb food, shows behaviour and engages in self-replicating activity is created successfully in a laboratory, as ultimately it no doubt will be, the momentous event will not shut down the debate over how life actually did form on this planet.*

Scientists have found that there are certain strands of nucleic acids, similar to those found in the DNA of modern cells, which will spontaneously assemble new small strands of nucleic acids, provided the right conditions are present.

These strands of nucleic acids, which are known as RNA, can carry genetic information that tells a living cell how to organize itself in very much the same way that DNA does today. However, researchers working with these RNA strands do not think they were just being used as instruction manuals on how to build cells in the days of early life. Instead, experiments suggest that RNA was also responsible for reproducing itself, working like an enzyme, which is something that DNA cannot do.

In laboratory experiments, when two RNA strands are placed together inside a confined area, researchers observe that one strand will spontaneously use the other as a template to construct a third strand that is a copy of the template. This fascinating behaviour of RNA is called replication and may hold the key to the fascinating manner by which the earliest life forms reproduced more of themselves.

Researchers working with these strands have also discovered that when they confine the RNA inside a sphere of lipid, the sphere will grow as it encounters and absorbs smaller lipid structures in the surrounding solution. Finally, when the researchers shoved the RNA-carrying lipid spheres through thin holes the size of those that might be found in coastal volcanic rocks, they found that large spheres with replicating RNA inside would often split into smaller spheres, each carrying strands of RNA.

What is astonishing about this research is that the mere presence of RNA seems to guide the behaviour of the sphere it is in. Spheres without RNA have relaxed surfaces – they do not seek out and collect smaller spheres of lipids in order to increase their size. However, spheres with RNA strands inside them have surfaces that are much more tense. The increased tension leads these RNA-carrying spheres to compete actively for empty lipid spheres that are present in the surrounding solution.

The researchers doing this work say that their experiments provide an evolutionary explanation as to why all cells today contain RNA. In ancient waters, if lipid spheres containing RNA were competing with one another to absorb extra lipids from the water, the RNA-carrying spheres with strands that could replicate the fastest and generate the strongest surface tension would have collected the most lipids and divided more frequently. Given that these successful spheres would contain copies of the same RNA template, their unique RNA 'make-up' would have soon become the most common sort of sphere on the planet. The speed of replication may have been the first evolutionary pressure for living cells.

Scientists love big questions, and they do not get much bigger than those about the earliest days of life on Earth. Even when a living cell that is organized, able to absorb food, shows behaviour and engages in self-replicating activity is created successfully in a laboratory, as ultimately it no doubt will be, the momentous event will not shut down the debate over how life actually did form on this planet. Palaeontologists and biochemists will go on looking for clues that tell us just how life on this planet began, what the early forms of life were like, why and how one thing created circumstances that led to another and, by inference, how in the end we are here today.

But of one thing we can be sure: once life got going, it rapidly exploded in abundance and dramatically changed Earth forever.

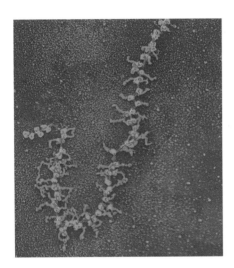

Ribonucleic acid, known as RNA, is composed of strands of nucleic acids which can carry genetic information and is thought to have been responsible for reproduction of the earliest life forms.

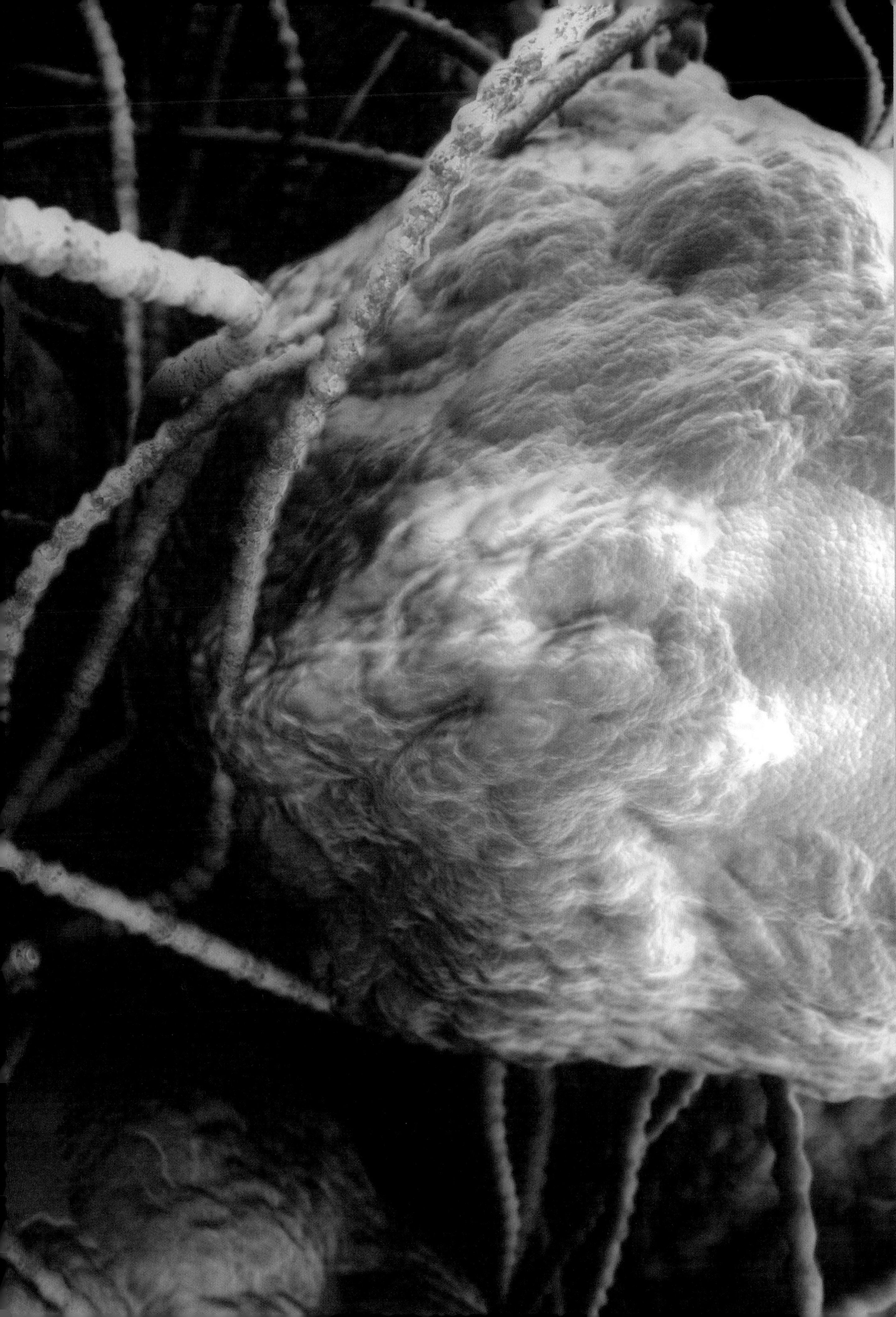

A RARE CALM has fallen over the present-day sea. The sun is warm, the water clear. We plunge into the swell and immediately enter a world of beauty and apparent peace, where the soundtrack of sea life hums along with precision. Here are pastel coral gardens and sudden rainbow flashes. Fish dart through the sunlight. Life on this reef is rich, varied and abundant. Often referred to as the rainforest of the sea, coral reefs may occupy less than 1 per cent of the world's ocean surface, but they form some of the most diverse ecosystems on our planet. Today, they are home to a quarter of all marine species.

For modern living things, including the species on the coral reef, there are two principal ways to feed. Plants gather a few nutrients from the soils that they live in, harvesting most of their energy from the sun and converting this energy into food. Animals and fungi, on the other hand, gather energy by feeding on other organisms.

By contrast, a 4-billion-year-old sea would seem barren and lifeless. The rich array of corals and fish that we observe living in our oceans today did not evolve until around 500 million years ago. But these waters in the very first oceans were still teeming with life – the microscopic life of early cells. But, unlike all modern living things, these earliest, single-celled organisms neither gathered energy from the sun nor fed on other organisms. They were fed solely by the volcanic nutrients released from the Earth into the surrounding water, and these nutrients kept the chemical reactions inside the cells going. At that time, there were no predators to eat the cells, and there was no need to search for food.

At some point in these ancient seas, an organism formed that could generate energy from exposure to the sun and live in waters that did not have all of the Earth-released nutrients that were normally required for early life to survive. This lifestyle change would have been the result of genetic mutation.

Mutation, in essence, is caused by a mistake being made during the copying of an organism's genetic information. When living things, including bacteria, reproduce, the nucleic acids (like DNA)

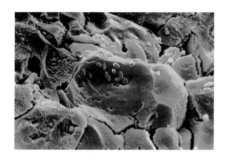

A scanning electron microscope picture of fossilized bacteria, which may be amongst the oldest evidence of life on Earth. These rod-shaped, red cells are around 3.5 billion years old.

that function as the genetic blueprints for their formation are copied. This copying process gives the newly formed organism a set of characteristics that will help it to become a fully functioning adult like its parent. However, sometimes this process goes awry and mistakes are made. Under such circumstances, the reproduced organism has slightly different nucleic acids in its genetic blueprints from its parent and therefore forms slightly differently as a result.

Mutations are common but most of them result in organisms with bodies that are either less effective or totally ineffective. However, every now and again, a mutation occurs that leaves an animal with a trait that actually helps it to survive. It is this concept of the evolution of a rare, helpful mutation that palaeontologists suspect played a major role in leading some early single-celled organisms towards a life that depended upon gathering some nutrients from the sun's rays. In fact, it might have been critical for the evolution of life.

Understanding how early life made the move from depending entirely upon the Earth's nutrients to depending, at least partially, on sunlight resolves only part of the dilemma. A critical question is why it migrated towards this existence in the first place. There is no way to be completely certain, but it is possible to produce a theory based on years of scientific study.

**Coral, seen here in shallow waters near the Great Barrier Reef, first evolved around 500 million years ago.**

# Understanding Precambrian Life

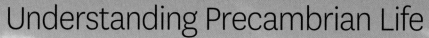

"What is very nice for me personally about *First Life* is that it really does complete my journey exploring life on Earth, although, of course, paradoxically it is the beginning of the story. It charts the arrival of life on this planet, which I've spent the better part of my time studying.

I have spent my career looking at animals in all stages of evolution. Although we have known for a long time details about life after the Cambrian period, what has been missing is the beginning of the story. We've always started at chapter two, thinking that there was no evidence before that. Chapter one was blank until around 50 years ago, when the first of the Precambrian fossils was discovered. Now this chapter has a considerable narrative, and it's continuing to expand.

Over these past 50 or so years, palaeontologists have accumulated enough evidence to put extraordinary detail into our understanding of how life evolved from tiny, unicellular life forms that still exist, like bacteria and the archaea, to creatures that became more complex.

Eventually, these creatures developed to the complexity of the small creatures that you would expect to find wriggling around at the bottom of a pond – flatworms, arthropods and tiny molluscs. These are quite similar to some of the ancient animals we find fossilized today, and it is these fossils that tell us how life began and developed.

When I started studying animal life, we had neither these very early fossils, nor an understanding of how this early life developed. So for me, it's very satisfying to be able to go back to chapter one."

If there had been a large number of single-celled organisms competing with one another for Earth-released nutrients in any part of the ancient oceans, such a scenario could have resulted in too many organisms and not enough nutrients in that particular area. This, in turn, would have led to many organisms dying off because they were not getting the required nutrients. Under such circumstances, any mutation that allowed for a simple organism to support its diet, at least partially, with nutrients from another energy source, such as the sun, would have had a tremendous advantage when it came to survival. This survival advantage would have allowed it to live while countless other organisms starved and, more importantly, it would have been able to pass along its genetic information to numerous offspring.

This process of a single individual appearing and being able to take advantage of its difference by thriving and multiplying is the basis for the entire idea of evolution. Had resources not been scarce, the mutation might not have had any beneficial effect for the cell or, worse, might have made the cell less effective at feeding on compounds drifting in the water.

However, if resources were limited and the mutation helped the cell gain nutrients that kept it alive, the mutant organism would have survived where others could not. We know this process as natural selection. It occurs when cells or organisms survive difficult situations through the use of a unique characteristic; the survivor then passes this unique characteristic to offspring so that it becomes more common and spreads through the population. It is a pivotal process that will come up again and again with more complex animal life.

————

Where is the evidence for this story line of early life? Palaeontologists studying recently extinct species such as sabre-toothed tigers, mammoths and even dinosaurs look at modern animals to develop theories as to how these ancient species might have behaved. Nevertheless, fossils are the cornerstones of palaeontology because they are a solid form of proof that a specific organism existed at a specific time in the Earth's history. Without fossils, it is impossible to know conclusively if and when an organism was alive. It is for this reason that the very earliest fossils are so extremely important.

The Earth is an active place, and the ongoing formation of rocks allows animal carcasses to become fossilized and be preserved. Since the geology of our planet has changed so much during the Earth's 4.5-billion-year history, there are only a few outcroppings on the planet where very ancient rocks are still present, and fewer still that were formed in such a way that the presence of early life was recorded. These rocks, ranging from 3.5 to 3 billion years old, are found in Australia and South Africa. They do not contain fossilized bones or even the fossilized impressions of ancient organisms because life 3.5 billion years ago had neither bones nor bodies big enough to be readily recorded by sediment. Instead, the fossils found within these rocks reveal the activities of ancient life.

Such fossils could be created today if you were to take a walk across a muddy field. You squelch through the mud and, when you reach the hedgerow on the far side, you look back and your footprints are clearly visible. Now it so happens that, after your little jaunt, the sun comes out and heralds an extended period of hot weather. The footprints dry and harden. Then, after hardening in the heat, they are covered in wind-blown sand or a new layer of mud brought into the area by a sudden flood. Your footprints are preserved under the new protective sediment. Left undisturbed, they stand a good chance of becoming fossilized over time. These footprints would become fossils that represent activity, even though they do not contain bones or full body impressions.

Trace fossils, as palaeontologists call them, can reveal more than footprints and movement. If a bird or an alligator were to build a nest and the nest were somehow to be preserved in the fossil record, that, too, would be a trace fossil. Even such unpleasant things as dung and vomit can, under the right conditions, be preserved as trace fossils. Indeed, dinosaur dung is a hot commodity among researchers who analyse it in order to better understand dinosaur diet.

Many Australian and South African fossils are the trace fossils of bacteria that were washed over by sediment billions of years ago. How do we know that these unremarkable piles of silt are actually fossils? Because structures that look identical to those found in the fossil record are still around today. One of the best places to find them is at Hamelin Pool in Shark Bay, Australia.

Shark Bay is a beautiful UNESCO World Heritage site in a remote area north of Perth. It is home to a large population of marine mammals called dugongs, as well as bottlenose dolphins and many threatened animals. The water is warm and shallow, but saltier than normal sea water. In the hot, dry air, the water of the bay evaporates fast, but large swathes of sea grass reduce tidal flows so the salt concentration remains high.

It is this extra-salty water that makes Hamelin Pool, located to the south of Shark Bay, such an exciting place and, more importantly, allows it to contain evidence of the beginnings of life on Earth. At first glance, a visitor will notice the mass of black, irregular rocks filling the shallows. A closer examination reveals some peculiar shapes, more like muddy rock columns growing from the sand, widening slightly on the way up to a flattened, dark, toadstool top. But these are not really rocks. They are the building blocks of bacteria and almost exactly resemble trace fossils dating back 3.5 billion years.

Organisms such as the cyanobacteria discolouring the water in this lake are thought to have helped change the early atmosphere by performing oxygenic photosynthesis.

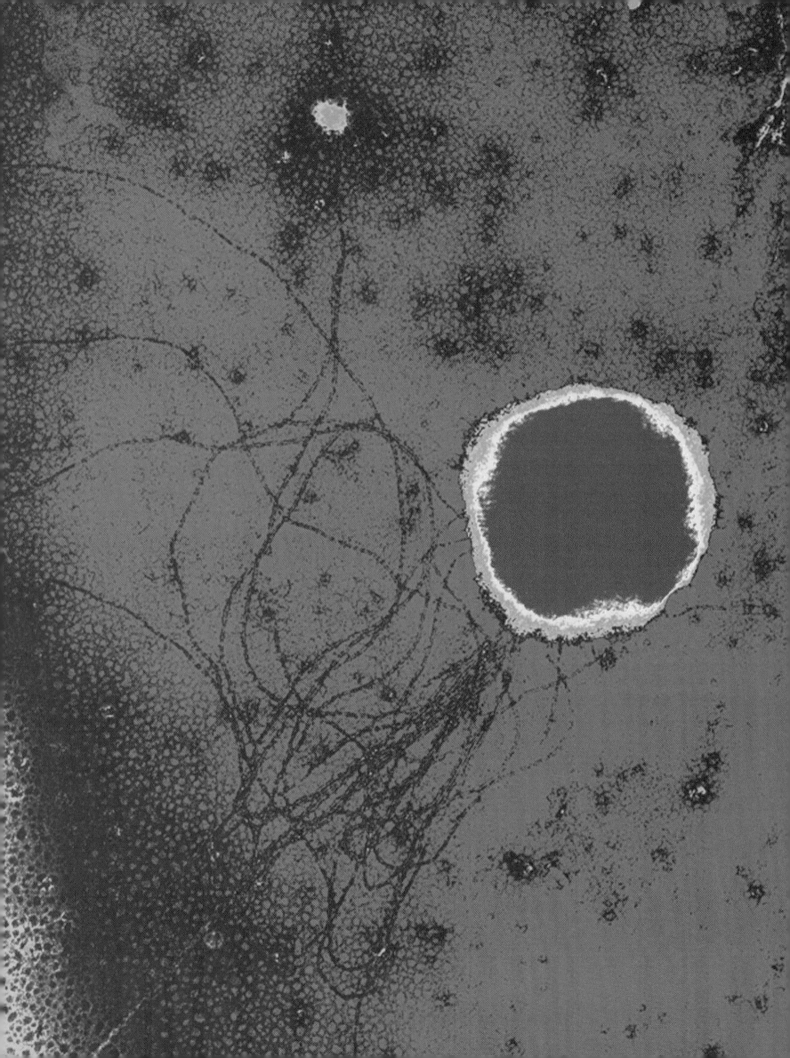

# The Power of Simplicity

" When discussing the evolution of life, we humans often hold the rather egocentric view that the closer something comes to a human in the evolutionary tree, the better or more advanced it is. By this reasoning, multicellular is better than single-celled, fish are better than insects, and mammals are better than reptiles.

I think you only have to observe the debilitating effects of the common cold or food poisoning on a human to realize that in fact micro-organisms often have the upper hand, and could be considered some of the most successful organisms on Earth.

Micro-organisms are the only form of life that has persisted since almost the beginning of life, and as such are living fossils. I got to experience some of these very early life forms for the first time around 30 years ago. These organisms can be found in few places on Earth, the most accessible of which is Hamelin Pool on the west coast of Australia.

The shallow, salty water of this bay is crowded with curious rock-like pillars called stromatolites, structures composed of the bodies and secretions of ancient photosynthetic bacteria. These are not dead relics, however. Populations of photosynthetic bacteria continue to replenish the columns, competing amongst one another and providing food for other, secondary bacteria that live in the lower layers. It's a bit like an apartment block for microbes, versions of which have been present on Earth for over 3.5 billion years. It's a concept that is hard to grasp, and I was struck with wonder at seeing such ancient organisms firsthand.

These life forms are so remote from our own, but they are life all the same. The truth is that all organisms differ in some ways. Some are certainly more complex than others, might have larger bodies or may be more successful in terms of their longevity or distribution on Earth, but it would be a huge folly for humans to think of themselves as the Earth's most significant life forms."

**Extremophiles, seen here (left) through a scanning electron microscope are photosynthetic bacteria which survived the Snowball Earth period which almost extinguished life. They live in the depths of the ocean at a temperature of 120 °C. These types of bacteria are thought to be very similar to the earliest bacteria, such as the fossil in Gunflint Chert (right).**

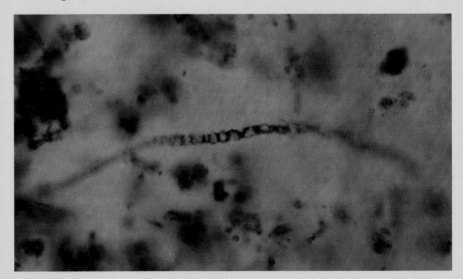

In Hamelin Pool, the saltiness of the water prevents would-be predators, animals like snails and urchins that feed on bacteria, from being able to enter it. In their absence, bacteria grow unhindered in a remarkable way. They need to be constantly exposed to sea water to avoid drying out, but they must also remain close to the surface of the water because they function like plants and draw their energy from the sun. For this reason, they are found in the tidal zone of the bay, where the water steadily rises and recedes. The bacteria have slimy protective layers around them and, over time, bits of sediment settle and stick to these layers. The bacteria respond to this by oozing their way up on top of the sediment and, in so doing, lock in sediment with their adhesive-like slime. This process of sediment accumulation leads the bacteria to create mounds with loads of trapped sediment behind them. But these mounds are alive. The bacteria at the top of the mound collect the visible light of the sun, but behind them are different bacterial species that collect the ultraviolet light, which can travel deeper through the surface bacterial layer. Then, lower down the bacterial pile, cells that do not feed on the sun's light feed instead on the remains of the upper bacteria that have died.

It is easy enough to mistake the mounds for the natural rock of Hamelin Pool. But these elaborate, slow-growing bacterial apartment buildings have been home to bacteria for several thousand years. They are called stromatolites by palaeontologists. Remarkably, they also appear in the fossil record and provide the first hard evidence of bacterial life 3.5 billion years ago. Whether the fossilized bacterial apartments found in Australia and South Africa originally contained the same bacterial species or different species from those found today is difficult to determine. But the fact that the stromatolite structures in South Africa look so similar to those in Shark Bay today suggests that if the bacteria were not identical, they were likely engaging in similar activities. Today, Hamelin Pool is one of only two places in the world that has living marine stromatolites, and it is the only place where they can be seen from the shore. They were discovered in 1956 by surveyors searching for oil and then examined by scientists. These stromatolites were the first-ever recorded evidence of living structures in Precambrian times. Though they first colonized there just 2,000–3,000 years ago, their lineage dates back 3.5 billion years to the dawn of life. For the next 3 billion years – until 500 million years ago – stromatolites were the only evidence of macroscopic life on Earth.

Just as modern apartments have limited space and can house only a certain number of residents, stromatolites as structures have limits, too. Bacteria, small as they are, do take up space. When the first stromatolites were forming, there were probably no predators, and there was likely to have been little competition for space. Sunlight was

These stromatolites in Australia are rocky deposits laid down by photosynthesizing cells, which are very similar to fossilized forms of the earliest ancestors of plants from 3.5 billion years ago.

abundant, there were numerous shallow coastlines all over the world for bacteria to build up communities, and life was in harmony. Still, paradise would not last forever.

As stromatolite communities spread, there would have come a time when ideal habitats filled up and competition developed. As new bacteria formed in this crowded world, many would find no space on the surface of the stromatolite to collect energy from sunlight and so they starved to death.

With space and access to sunlight in the stromatolites becoming increasingly limited as bacterial populations grew, any bacterium forming that could exist with a fraction less sunlight than the rest of its kin (or had some way of making a living that did not involve sun exposure) would have been able to survive. Mutation is, again, thought to have been the way that life moved forward.

In modern stromatolites, there are bacteria that live beneath the surface layer of bacteria and make use of ultraviolet light passing through the upper layer. Because they are surviving on a different type of light that penetrates deeper into the stromatolite, these bacteria do not have to occupy the surface of the stromatolite structure. It is possible that bacteria similar to the modern ultraviolet-collecting bacteria arose as a result of a mutation as the surfaces of stromatolites became more and more crowded. Even a minor mutation that allowed the slightest ability to absorb energy from ultraviolet, rather than visible, light would have been a boon to these organisms because it would have prevented them from having to compete for surface regions of the stromatolite.

Today, there are other bacteria in stromatolites that are specialists at consuming dead and dying light-collecting bacteria. These organisms, which, in effect, are scavengers, could also have arisen as a result of mutation and been driven by natural selection to take on this unique feeding role.

The rocks in which trace fossils are found also provide vital clues about the alien world of these bacteria. Even the atmosphere was totally different. Today, we take the presence of oxygen in the air for granted. We are constantly breathing in oxygen and exhaling carbon dioxide. Dogs, cats, elephants and whales all do exactly the same. It is only when we go to places where we cannot breathe – in space or under water – that we must take air, containing that vital oxygen, with us. Plants effectively function in the reverse. They breathe in carbon dioxide and exhale oxygen – a good thing, too, because without plants the planet would be overwhelmed by carbon dioxide, and all animals would die. Intriguingly, 3.5 billion years ago, as life was just getting started, oxygen was almost entirely absent from the atmosphere. While it accounts for about 21 per cent of the gas in the atmosphere today (with nitrogen making up most of the other 79 per cent), the chemistry of rocks that formed during life's early days suggests that oxygen made up less than 1 per cent of the gas on the planet.

This is not surprising. None of the chemistry experiments that have explored the creation of life from early Earth compounds involves oxygen. Indeed, if oxygen were to be included in many of these experiments, it would damage critical compounds thought to have been needed for life's evolution. Yet, some very strange geological formations – blood-red, wavy bands of iron – 2.5 billion years ago suggest that conditions on Earth changed. In spite of their extreme age, they are readily found around the world. Known by geologists as banded-iron formations, they are commercially very important and supply much of the iron that the world depends on for construction. They are also an enigma. Banded-iron formations appear to have developed at the bottom of the ocean between 2.5 billion and 1.8 billion years ago, but they did not accumulate steadily. The wavy bands show periods of sudden formation followed by periods of little or no accumulation, leaving geologists to work out what was happening.

—

*While oxygen accounts for about 21 per cent of the gas in the atmosphere today, the chemistry of rocks that formed during life's early days suggests that oxygen made up less than 1 per cent of the gas on the planet.*

One theory is that photosynthesis, the chemical process that plants today use to collect energy from the sun, became extremely common around the Earth at the time when the banded-iron formations started coalescing. While there were most certainly no actual plants present 2.5 billion years ago (we would find fossils of their fibrous stems and leaves if they had been around), there were plenty of single-celled organisms, such as the bacteria dwelling in stromatolites, and blue-green bacteria that still live in many aquatic environments today, that would have been engaging in photosynthesis. Through photosynthesis, stromatolites produce carbohydrates from carbon dioxide and water, a process that releases oxygen into the air.

Strangely, geologists do not see signatures in ancient rocks showing oxygen levels reaching the 21 per cent level that we see today. Indeed, geologists do not even see oxygen-level signatures in ancient rocks creep much above 1 per cent until 2 billion years ago, and levels seem to hit the 20 per cent mark only around 500 million years ago. The banded-iron formations present an explanation of what was happening to the oxygen in a world where early photosynthesizing organisms exhaled it, but where no animals were around to inhale it.

Today, iron is thought of as a metal found among rocks and not easily dissolved in liquids. However, in the absence of oxygen, iron readily dissolves in water. On the early Earth, the oceans were probably rich in dissolved iron since no oxygen was present to prevent the dissolution. As photosynthesizing bacterial populations increased through the ages, the oxygen released most likely reacted with iron that was dissolved in sea water. The iron turned to rust. As populations of ancient photosynthesizing, single-celled organisms increased, tiny flakes of rust-red, oxidized iron are thought to have appeared, gently falling and settling on the ocean depths.

But why the iron stripes in the rock formations? In modern lakes and oceans, provided that there are enough nutrients in the water and the weather conditions are just right, plankton populations can explode. If such population explosions took place periodically in the past – perhaps the result of a rush of nutrients flowing into the oceans from rivers and/or periods of perfect plankton weather – the tiny cells would together have released vast quantities of oxygen in a burst. These sudden releases of oxygen would have led to an abundance of oxygen quickly interacting with dissolved iron, causing large amounts of it to fall through the ocean water and create distinct layers, or bands, of iron on the ocean floor.

Over time, these bands of accumulating iron would have come under pressure as more and more flakes of rust fell down, creating piles that were many kilometres thick. Ultimately, the pressure created by kilometres of iron-rich sediments would have transformed the bands into the solidified banded-iron rocks. Most of these rocks have been destroyed over the ages as our restless planet has melted and recycled them but, even so, there are so many surviving kilometre-thick banded-iron formations today that they must have been extremely common long ago.

——

Between 2 billion and 1.5 billion years ago, the plant kingdom was in full swing perfecting the art of photosynthesis. Red rust was falling to the sea beds and, over the ages, oxygen was enriching the atmosphere. And as the conditions changed, so did life – in a subtle but, nevertheless, important, way.

Multicellular plants and animals did not exist yet, but the single-celled organisms on the planet were fulfilling all of the same roles that we see in the natural world today. The situation was effectively the same as having plants that collect energy from the sun, insects that eat the plants, and birds that eat the insects. It was just all taking place on a microscopic scale.

However, at some point this simple world became more complex because individual cells started to collaborate with each other. Palaeontologists have determined that between 2 billion and 1.5 billion years ago, single-celled organisms began consuming other single-celled

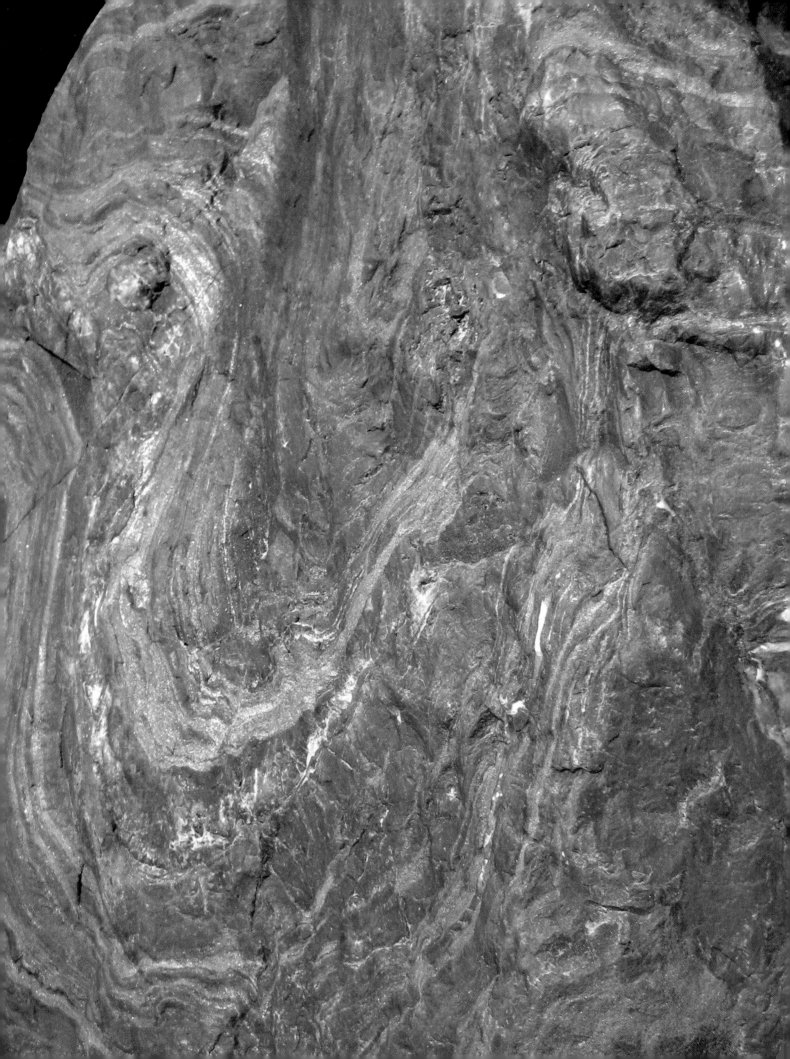

organisms, but instead of digesting and destroying them, they captured their prey alive and engulfed it whole without killing it. However, it is suggested that when the first cells began this behaviour, it was most likely unintentional and the result of a mutation that caused the cells to engulf but not kill other cells – after all, what good is a predatory cell if it can't kill other cells? This change in behaviour may simply have been a useful mistake.

Exactly what the mutation was is much debated. It is argued that perhaps the membrane surrounding the predatory cell was more flexible than the membranes of most other single-celled organisms at the time. Or it could be that the predatory cell developed a mutation that allowed it to grow a bit bigger than its kin and thus absorb, rather than destroy, its prey. It is also argued that single cells ended up inside other cells as the result of cellular division gone awry.

Regardless of the mechanism, the results are visible in living cells today. Locked inside animal cells, including those within our own bodies, there are tiny biological mechanisms, or organelles as they are known, such as mitochondria. These organelles consume nutrients and release an energy-packed compound that cells use to power their day-to-day existence. The mitochondria live entirely within animal cells but they are different from all the other bits inside – they actually

Left: Banded iron formations provide clues to changing conditions on Earth around 2.5 billion years ago.

The presence of DNA strands in mitochondria, organelles which live within animal cells, suggests that they were once single-celled organisms in their own right.

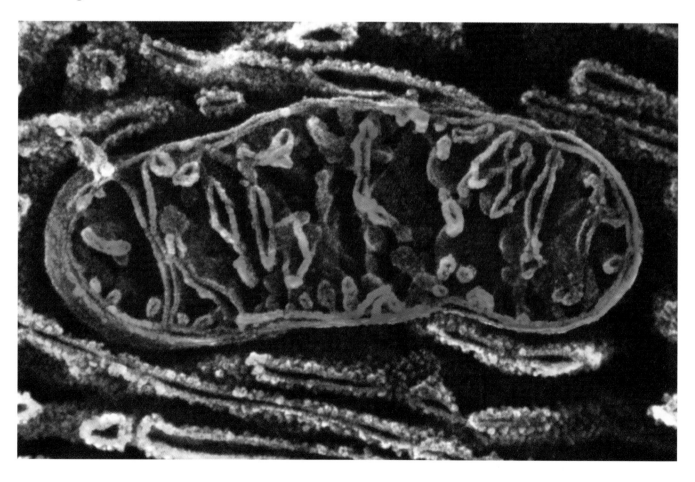

carry some of their own genetic information in the form of tiny strands of DNA that are different from the strands found in the host cell.

The presence of DNA strands in mitochondria suggests that they have not always been organelles within other cells but were once individual single-celled organisms in their own right that were somehow absorbed by ancient predatory cells.

Mitochondria are not alone in their status as single-celled organisms that were collected by larger cells. Chloroplasts are the green-coloured organelles found inside plant cells. These transform sunlight into energy-packed compounds that plants use for growth and reproduction. Like mitochondria, chloroplasts have their own genetic information. In addition, they bear a striking resemblance to very simple bacteria that use energy from the sun's rays to fuel themselves. For these two reasons, researchers believe that chloroplasts were once sunlight-feeding, single-celled organisms that were later absorbed or collected – but not digested or destroyed – by larger single-celled organisms.

Such absorption scenarios sound a bit like enslavement and a raw deal for the ancient ancestors of modern mitochondria and chloroplasts. But, in reality, the situations probably benefited both the captors and the captured. The captured cells were granted an extra level of protection, while the captors were provided with nutrients and energy that resulted from the captured cells' activities.

The fact that such collaboration took place is not surprising. The natural world is full of situations where two species interact in ways that are beneficial to both. Back under the ocean in the warm coral reef in our world, there's an example of this in the form of a fish cleaning station. Here, a slender bluestreak cleaner wrasse, with its distinctively dark mouth-to-tail stripe, advertises its services to passing shoals. Those fish needing a scrub stiffen and let the wrasse get to work. The wrasse feeds off the parasites on the skin of the other (often larger and predatory) fish species. The 'client' fish does not eat the wrasse because it wants a good clean; the wrasse gets both protection and a good meal from the arrangement.

———

Even more similar is the collaboration that exists with our gut bacteria. Inside the human gut are swarms of bacteria that are not technically part of our body. They feed on the food that we eat. By doing so, they break down food into digestible bits that the body can use. It is a partnership that benefits both us and the bacteria. Different species of 'good' bacteria are found in just about all animals alive today.

Clearly, collaboration is a characteristic that emerges naturally among living things, but as early single-celled organisms were engaging in collaborations by absorbing one another, a new collaboration was taking shape in the form of innovative reproduction.

Bacteria today engage in a behaviour called conjugation, where one bacterium injects a bit of its genetic information (a few of the nucleic acids that make up DNA) into another bacterium. It is an exceptionally useful behaviour because when rare, helpful mutations are made as the nucleic acids copy themselves, these beneficial mutations are not just passed along to the offspring of the bacterium in which they formed, they can also be passed along to other unrelated bacteria. Through the regular sharing of genetic information, bacteria are able to make communal use of mutations that accumulate in a population over time.

This is very good for the bacteria. It allows them to quickly share immunity to antibiotics when they encounter them in hospital settings and survive conditions that would otherwise kill them. Of course, this is something of a problem for humans, as it leads to antibiotic resistance, MRSA and 'superbugs'. If one bacterium, through mutation, develops a resistance to an antibiotic, the information for this resistance perpetuates. Eventually, bacteria become immune to more and more antibiotics, and they become so hardy that they are nearly impossible to control when they infect the human body. Indeed, while conjugation is great for bacteria, it makes them extremely difficult to manage medically, which, ultimately, is very dangerous for humans.

Exactly when this behaviour first occurred is a mystery. Tiny bacterial cells are incredibly hard to find in the fossil record. Finding evidence for them injecting bits of their genetic information into one another is virtually impossible. Yet, because they do it today,

Chloroplasts convert sunlight to energy. Very similar to mitochondria, they are also thought to have originally been free-living organisms.

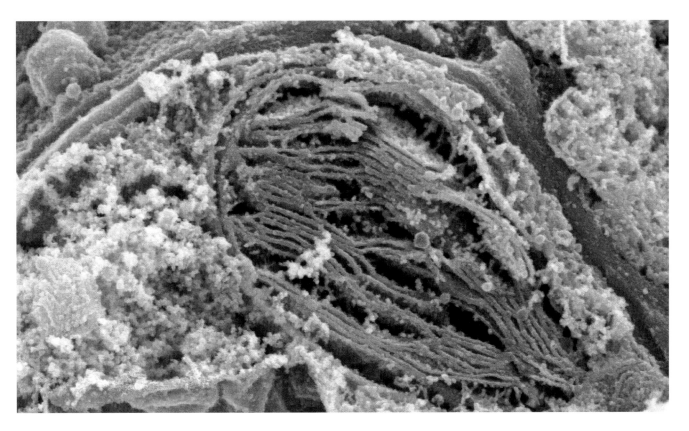

they probably did it in the past. Some palaeontologists speculate that by the time life on Earth started to become more complex between 2 billion and 1.5 billion years ago, conjugation was fuelling the evolution taking place. However, it is also possible that the increasingly rapid changes in primitive life and the rise of more complex organisms were the result of something else. It is possible that sexual reproduction was evolving.

For life to survive, it has to keep changing because the world around it is constantly changing. What sexual reproduction does is allow species to change by shuffling the genetic deck. Therefore, every time offspring are produced, a new individual with genetically new characteristics is created. These offspring are evolutionary experiments, so some of them die out and others survive. The shuffling means that only the stronger ones carry on to a point where they, too, can reproduce to create their own genetically different (and usually genetically stronger) offspring.

In the natural world, sex is, at its core, an investment strategy. Before the advent of sex, when it was time to reproduce, all organisms cloned themselves. Cloning is the process in which an organism duplicates its genetic material through asexual reproduction. It is the quick and easy route for reproduction: no need for a mate and with little investment of energy or cellular resources.

Clones, however, are generally bad for long-term survival because if, for example, the climate changes or a disease strikes and the organism is vulnerable, then all members of the population will simultaneously die. Clones have no genetic diversity and, therefore, all members of the group share the same vulnerabilities. This is why, initially, antibiotics are so effective against bacteria that infest the body.

Sexual reproduction allows two organisms to create variety in the population and avoid the inherent insecurity of cloning. The process of sex involves each organism donating genetic information. This information is mixed together and roughly 50 per cent of the information from each of the two parent organisms is passed along to the offspring that are ultimately created from the genetic mixing.

When sexual reproduction first evolved among clones, it would not have amounted to much since the clones would mix identical genetic information together. However, as random mutations built up in the population, sex would have allowed the genetic information responsible for the mutations to be shuffled, thereby creating new mutant variations. Then, when these variant organisms later reproduced through sex, they, too, would have shuffled their genetic information and created even more diversity. This genetic shuffling and mixing from two reproducing individuals is why children often resemble both of their parents and sometimes even have characteristics that seem to belong to both of them.

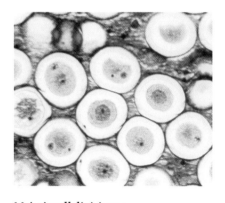

Meiotic cell division to produce germ cells, seen here in roundworm ova, is one of the key stages in sexual reproduction.

Sex generates variety; it is an insurance policy for the genetic information inside organisms. As an example, let us drop into the sun-dappled world of the coral reef. A shoal of Spanish mackerel, all average-sized, twists and turns and catches the light on their silvery scales. If the shoal were composed entirely of clones, and if the temperature of the water suddenly rose enough to prove dangerous to the mackerel's health, all would die at roughly the same time. The same would be true if a new predator, perhaps a white-tipped shark, came into the region and was just fast enough to chase after, and eat, the mackerel. It would soon catch them all since they would all swim at the same speed.

Fortunately for mackerel, they do not clone themselves. Instead, they reproduce sexually and their populations have a lot of variety. There are some big, slow mackerel, some average-sized, and some that are lean and particularly fast. With such variety, the population responds and develops according to changes in the world around it.

*Through the regular sharing of genetic information, bacteria are able to quickly share immunity to antibiotics when they encounter them in hospital settings and survive conditions that would otherwise kill them.*

If changes in sea currents exposed these fish to much colder water, the average-sized and lean fish may struggle to adapt to their changing habitat and die out, but the bigger fish would be more likely to survive. This would mean that the mackerel doing the bulk of the breeding would be the surviving larger, slower ones. Since these big and slow mackerel would be carrying genetic information that would lead to the birth of big and slow offspring, the population, rather than collapsing, would evolve from being a mixed population of lean, average and big mackerel into a predominantly big and slow population.

But what if a white-tipped shark came along that was much faster than the average-sized mackerel and the big, slow ones but not quite fast enough to catch the smaller, lean fish? It would be the lean and fast mackerel that would survive and breed. This exclusive breeding of lean and fast individuals would lead to genetic information coding for lean and fast body forms to dominate in the mackerel, and the population would evolve into one of leaner and faster animals.

So, sex is a powerful protection for genetic information inside organisms. For a population and its genetic information to be wiped out completely, organisms need to experience extraordinarily bad luck. This could be a sudden change in ocean currents that occurs at the

same time as the appearance of an aggressive predator. Such a scenario, while rare, is referred to as an evolutionary trap because of the way it corners and destroys populations.

But there is a price to pay with any insurance policy. When organisms reproduce by creating clones of themselves, 100 per cent of their genetic information is passed along to their offspring. This means that they are effectively perpetuating all of their genetic information for an entire new generation by simply reproducing once. For an organism that reproduces using sex, only about half of the genetic information is passed to the next generation, with the other half coming from the partner. This means that sexually reproducing animals need to create twice as many offspring to pass along their unique genetics to the next generation – an extreme demand.

Even so, the adaptability that sex provided organisms by allowing lineages to flex rather than collapse under stressful circumstances probably led it to spreading rapidly as a means of reproduction. Unfortunately, like conjugation, it is impossible to know for sure when sexual reproduction arose but, again, researchers suspect that the appearance of many new single-celled organisms in the fossil record around 1.5 billion years ago is a hint that sexual reproduction had been invented. Without it, the evolution of life could not have proceeded.

As busy as Earth was, relatively simple microscopic organisms were, for some 3 billion years, the most advanced forms of life on the planet. They provide incredible insight into life at those almost incomprehensibly early stages and into the evolution of traits necessary for survival. But, suddenly, within the space of a few million years, a mere blink of an eye in evolutionary terms, advanced organisms appeared. Why is a mystery, but the search for answers starts as the Earth is plunging into the coldest, longest winter imaginable.

**A group of fossils preserved together such as these trilobites found in Morocco is called a 'life assemblage' because it shows a group of animals that were killed suddenly and then preserved where they lay.**

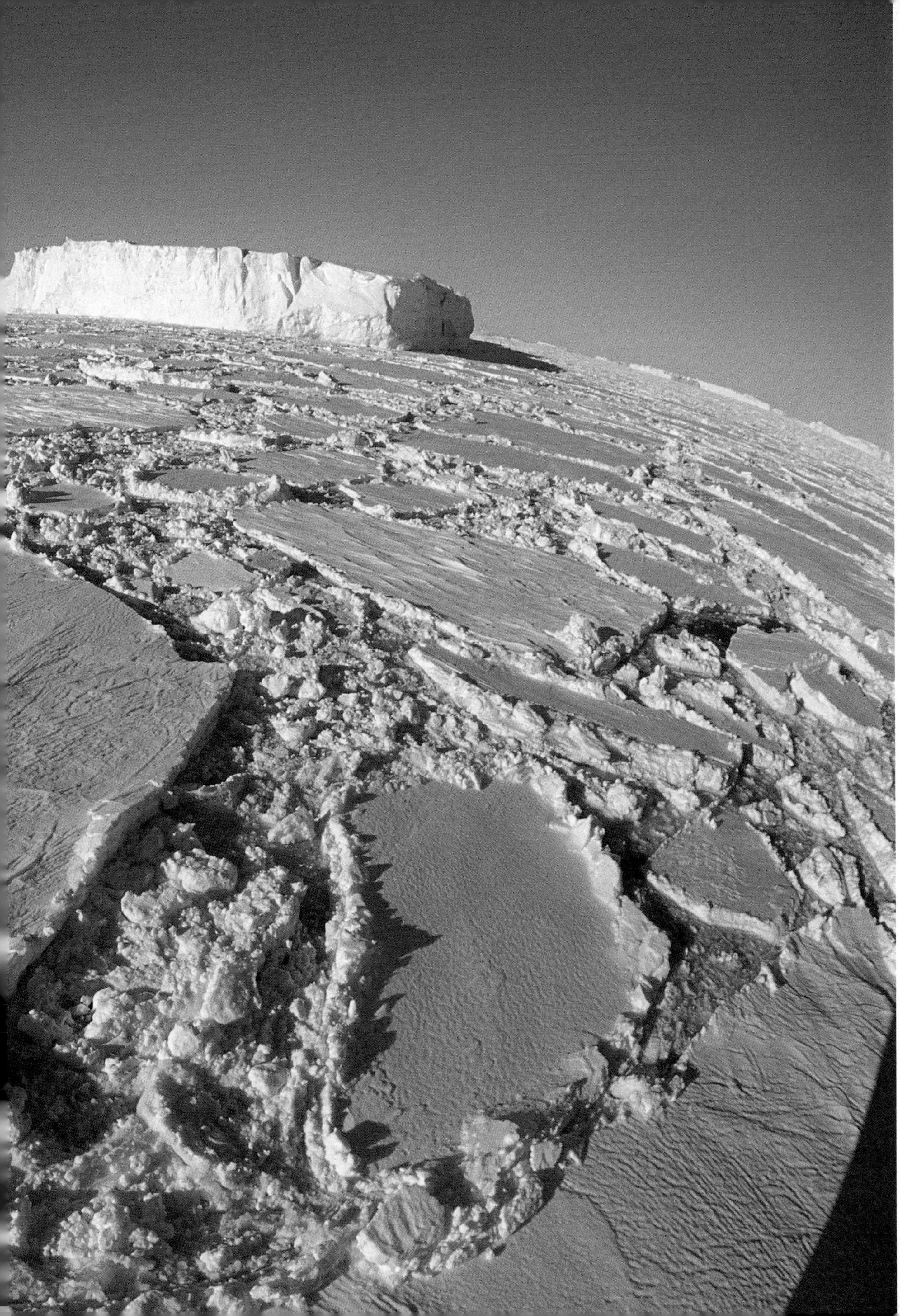

# CHAPTER 3 • SNOWBALL EARTH

HIGH IN THE Canadian Rockies, heavy snow begins to fall. This is a full-on blizzard, an extreme mountain snowstorm of flying ice created by plummeting temperatures and whipping winds, and poses a serious challenge to scientists like Professor Hazel Barton.

A microbiologist at Northern Kentucky University, Barton spends much of her time in the lab peering through microscopes at blobs – translucent bacteria – on sample slides. But up here in the Columbia Icefield, she is prepared for these conditions. The sudden squall may seem vicious, but she knows this mountain well. It's a feint, she's sure, and she is determined to reach the glacier. So she pulls off her backpack, takes out some chocolate from her stash and waits for the storm to pass.

The sharp sting of ice crystals in the air is a reminder of what conditions on Earth were like in the past, a chill that would have been felt across the entire planet. Scientists believe that about 700 million years

ago, the Earth, for reasons that are not fully understood, was transformed from a blue planet into a massive snowball. It was a 100-million-year period of glaciers advancing, building pockets of ice and then retreating. Though this period of glaciation probably wasn't a total freezing of the planet, it was a severe ice age that irrevocably changed the living world.

But could life survive for millions of years at temperatures that may have dipped as low as -40° C (-40° F)? It seems unlikely. Does the evidence stack up? How geologists pieced together the puzzle of the period known as Snowball Earth over the past hundred years is a wonderful reminder that scientific discovery is sometimes far more like a Sherlock Holmes adventure – hours spent debunking impractical theories – than eureka moments occurring in sterile labs. Some think it might be a coincidence that animals appeared after Snowball Earth, while others put forward a compelling argument that this epic freezing of the planet was the trigger for the evolution of animal life.

It is hard for palaeontologists to look back through the geological record to prove the extent and duration of Earth's ancient frozen past. The history of volcanoes can be easily traced because they leave lots of evidence. Through the lava and ash that they eject, traces of the most ancient eruptions can be seen in the rock layers. Big storms also leave evidence by creating floodwaters that deposit large amounts of debris that ultimately get buried and recorded in sedimentary layers. But ice? It just forms slowly and then melts into water when temperatures warm up. Isn't an icicle the perfect murder weapon that would thwart even Holmes's mythical detective powers?

But even ice leaves clues. As the sun comes out again in the ice fields in the Rockies, Hazel Barton can see all around her how ice is changing the landscape; these same powerful processes were at work hundreds of millions of years ago, and it is the trail of destruction left by the ice that has led scientists to assert that for a period of around 100 million years ice covered the surface of the Earth. 'It looks like everything's been wiped clean,' Barton explains as she looks out across the ice field. 'The glacier's come through and destroyed all life. There's nothing living, but to a microbiologist this looks a bit like a rainforest. The discoloration on the surface of the ice is not dirt. It is bacteria that's surviving there, and that creates an ecosystem just like a rainforest, where you have plants, and you have predators that come in and feed on those organisms.'

In the Columbia Icefield, and in mountain ranges all over the world, there are areas where ice exists year-round because the weather never gets warm enough to melt it completely. Known as glaciers, some are rounded ice forms that sit on the sides of high-altitude peaks; others are long rivers of ice that run for miles.

The Columbia Icefield sits between several high peaks in the Rockies between Banff and Jasper. It resembles a giant saucer of ice, and, as the snow falls each year, the saucer fills up and ice spills over

The Canadian Rockies still hold important clues about early life for microbiologists.

# Dropstones

" Some of the rocks on the Avalon Peninsula coastline in Newfoundland, Canada, which we visited while filming *First Life*, have been shown by radiometry to predate the time when life became more complex. If we can understand the circumstances under which these rocks were laid down, then we can begin to comprehend why life suddenly became very complex.

Fragments of red stone are imbedded in some of the rocks. They look out of place – and they are. These dropstones were carried many miles by an ancient glacier, ground up and smoothed off by the intense pressure of the ice, and then deposited in sediment at the bottom of the sea to be later incorporated into a sedimentary layer.

Canada is a place where encountering a glacier wouldn't be a surprise, but what about Australia? When explorer Sir Douglas Mawson found dropstones in Australian sedimentary rock it must have come as something of a surprise to him. Although the continents have shifted a bit since 700 million years ago, at that time Australia was positioned firmly in the tropics. So, if glaciers did form there, then the Earth must have been frozen as far as the Equator, or close by.

The idea of our planet bound in snow and ice almost defies the imagination, and indeed, for some scientists, the idea of a 'Snowball Earth' seems implausible. Great debate has surrounded this topic during my lifetime, with some highly respected scientists still refusing to believe. Personally, I was slightly sceptical about the idea when I first heard about it, but it's a theory I've come around to over time, having closely followed both sides of the argument."

Evidence, such as the dropstones found across the world, points to a worldwide spread of glaciation around 700 million years ago.

the rim. Eight ice rivers, including the easily accessible Athabasca Glacier, extend from the ice field and 'flow' down the steep mountainsides. They appear completely still, but these glaciers behave much like rivers, in extreme slow motion. New ice forms at the top of the glacier every year during cold-weather storms, and this exerts pressure on the lower sections of the glacier, slowly moving them downhill. They are literally flowing ice, and they cause major alterations to the landscape.

As it flows down from the mountains, the ice grinds against the rocks in its path, picking them up as it moves along. Small rocks get stuck in the bottom layers of the ice, while boulders from the cliffs along the sides of the valleys fall down on top of the ice and are carried by the slow-motion current. These rocks are the clues that geologists prize because, even if the ice melts away, the rocks - and the damage they cause as the glacier propels them slowly down the mountainsides - remain. For example, the small rocks and stones caught underneath the ice are dragged along as the glacier moves and act like sandpaper; over millions of years, they carve the land and leave behind marks that geologists can see.

Moving glaciers also collect large amounts of dust, which, along with the rocks, stones and boulders, get dumped at the end of the glacier. This occurs either where the ice meets the sea, which causes it to break apart, or at low elevations where temperatures are warm enough to cause it to melt.

On land, when the ends of glaciers melt, the stones carried by the glacier form a big pile. Geologists recognize these ancient dumping grounds. The debris from ancient glaciers looks exactly like the mounds of rock and dirt, called moraines, found at the foot of glaciers today. Indeed, as Barton hoists her backpack onto her back and continues her climb, she is walking across a vast glacial rubbish tip. The stones that crunch under her boots, and the mud that splashes up her trousers, reveal the work of ice. The stones show evidence of having been chewed up by the glacier, and they are unlike other stones from the area since they have been transported quite a distance from their original location.

At sea, the glacier leaves another form of obvious evidence. As it breaks apart, it forms icebergs that float away. These icebergs carry many of the stones that the glacier collected. As the icebergs melt, they drop the stones into the ocean, where they stand out as distinctly different from everything else.

Just like the formation of fossils, the stones dropped on land and sea can be covered by sediment that solidifies into rock. This solidification preserves the deposited stones, known as 'dropstones' by the glacial geologists who look for them. Dropstones provide excellent evidence of the presence of glaciers.

Aerial photograph of the folds of a glacial moraine on the Malaspina Glacier, Alaska. Moraine is rubble material that has been transported and deposited by a glacial surge.

Sir Douglas Mawson was among the first to climb the Antarctic volcano Mount Erebus, with Sir Ernest Shackleton in 1908, and he would have been well acquainted with glaciers and moraines. So when he stumbled upon a lot of dropstones in Australian rocks, he knew exactly what they were. But his discovery surprised him because Australia is rather hot today. Mawson theorized that for dropstones to be present in such a climate, there must once have been extensive glaciers. It was an interesting theory and one that planted the seed of an idea: glaciers might have formed in tropical latitudes. However, it would be decades before that seed would sprout into the theory of Snowball Earth.

In the years after Mawson's findings, more ancient dropstones were collected around the world. Evidence continued to mount that at some point between 720 million and 660 million years ago, the planet became very cold. If the dropstones had appeared only in areas of high latitude, near the poles, for example, the discovery of so many glacial deposits would not have stimulated much interest – geologists had collected clear evidence of the occurrence of more recent ice ages, in which big glaciers would have been common towards the North and South Poles. But a discovery by a geologist called W. Brian Harland suggested that the freeze of 700 million years ago was something special. He argued that glaciers must have also formed throughout the tropics.

Since the days of Mawson, geologists had been using chemical analysis to work out the age of dropstones to determine when they were deposited by thawing ice. But Dr Harland tried a new tactic not just to find out when they had been dropped, but also where. His findings proved to be the first hints of the extent of Snowball Earth.

———

At first glance, working out the location of a specific dropstone seems like a matter of geography. If the dropstone is dug up in the tropics, that means there must have been a glacier long ago in the tropics, right? This is certainly what Sir Douglas Mawson believed, but we now know better. With very old rocks, it is not so simple to determine where on the globe the rock was when it formed. As we saw in the first chapter, the ocean crust and the continents are restless, constantly moving at a speed of roughly 5 cm (2 in) a year. That does not sound like much, but over 700 million years, a continent travelling at that rate can move 35,000 miles – which is nearly one complete circuit around the surface of the Earth. So if, like Mawson, you find a 700-million-year-old dropstone in the dry, hot hills of southern Australia, you have no way of telling whether it was dropped in a tropical or polar location because the Australian landmass has moved around and spent plenty of time near the poles where glaciers were common.

Sir Douglas Mawson, the British-Australian geologist and Antarctic explorer whose work on glaciation would lead to the theory of Snowball Earth.

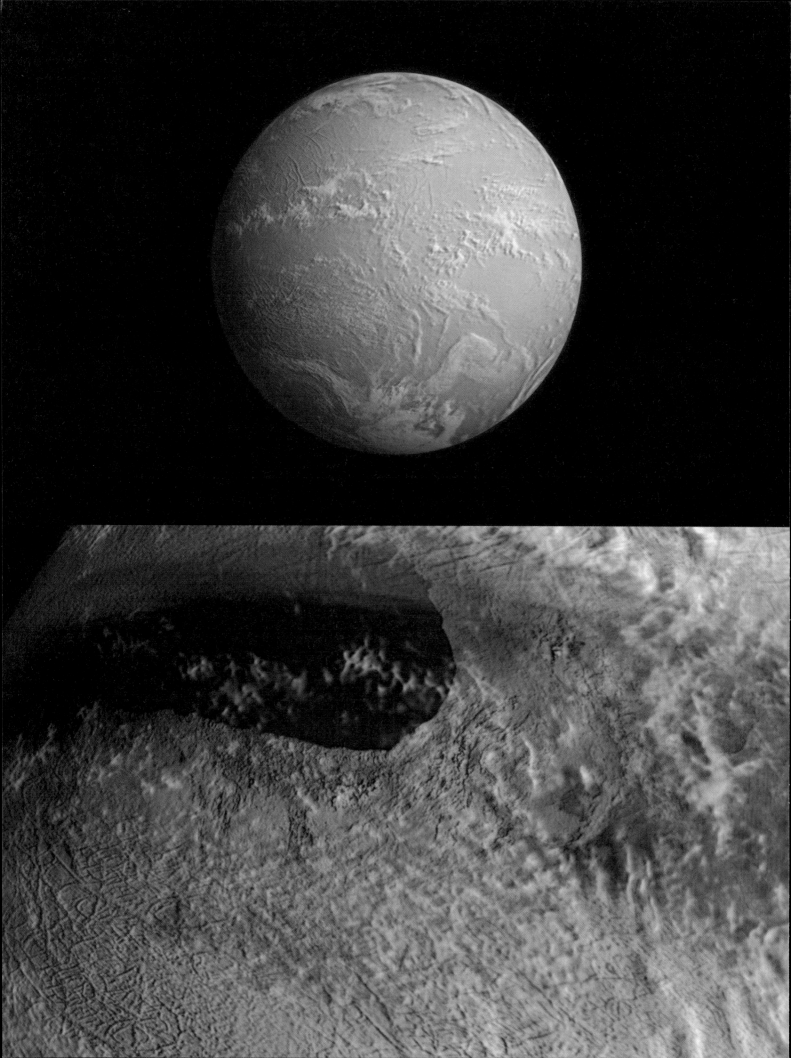

However, Dr Harland worked out a new analytical method that could determine where dropstones were originally deposited: palaeomagnetics. This relies on the fact that over time the Earth's magnetic field has changed in intensity and geographical position.

At its core, the Earth is thought to be a big, hot ball of iron. Since we know the planet spins around, this molten iron centre spins as well. Researchers know that if you quickly spin a giant piece of metal such as iron around, you create a magnetic field. So the Earth's rotation turns this planet into a gigantic magnetic field. Just like the magnetic ornaments stuck to the door of a refrigerator, the Earth has a North Pole and a South Pole, as well as a magnetic field.

It is the Earth's magnetic field that Hazel Barton's compass picks up as she takes a bearing on her route up to the glacier. Her compass needle aligns itself with the magnetic field, with one tip pointing towards the location where the magnetic field emerges on the surface of the planet, quite close to the North Pole. Once she knows which way is north, Barton can then plot her route on a map.

In the 1950s, scientists discovered that the Earth's magnetic field has not always emerged from the same location. Today, Barton's compass needle points north, but 800,000 years ago it would have pointed south, as the magnetic North Pole was somewhere in Antarctica. How do we know this? Well, of course, there were no geologists around millions of years ago to keep records but, rather remarkably, there were compasses, though not like those we are familiar with today.

When lava shoots out of a volcano, its minerals are in a liquid state. Some of these minerals contain iron and, when they cool down, they become magnetic. As these components settle, they align themselves to the magnetic field. By looking at iron-containing minerals like magnetite in ancient lavas that can be chemically dated, geologists have been able to work out that the Earth's magnetic pole often reverses itself from north to south and back again. The last reversal took place 780,000 years ago; before that time, there was a long period when magnetic north was actually in the south. Why the magnetic field does this is unknown, but the orientation of magnetic minerals in ancient rocks show that reversals take place every 300,000 years on average.

But palaeomagnetics is more than identifying magnetic reversals because you can use the orientation of minerals to find out where a rock was formed. If cliffs containing minerals rich in iron are blasted by wind and rain and pieces fall off as bits of sediment into a river, like molten rock that is cooling, these iron-rich pieces will orient themselves towards magnetic north. This is useful for geologists because sediments, unlike molten rock, often form in neat parallel rows, like the layers of a layer cake. (This happens because, when rivers enter calm water, they drop heavy sediment like pebbles first, then medium-weight sediment like sand, and, finally, light sediment, such

During Snowball Earth, the planet was plunged into a deep freeze so severe that it probably extended from pole to pole – although no one is sure of the extent of the freezing. It was probably a global surge in volcanic activity that brought Snowball Earth to an end.

as silt and mud.) When these layers transform into rock, they may be turned and flipped as the Earth moves around. Geologists can use the orientation of the layers to work out where the flat ground was when they formed. This knowledge provides those detective geologists with a crucial crumb of information: they can look at how the magnetic bits of sediment are oriented relative to the flat surface of the Earth.

Consider the image of the Earth's magnetic field. At the Poles, it is effectively emerging from the ground - your compass needle would try to point straight down at your toes if you stood on the magnetic North Pole. At the Equator, the field is effectively horizontal relative to the surface of the planet. This means that magnetically oriented minerals that settled in sedimentary rocks near the Poles should be oriented vertically, pointing down to the bottom of the layer cake of sediment. However, if the magnetic minerals settled in sedimentary rocks near the Equator, they would orient themselves parallel to the layers of the layer cake.

**Mono Lake, California, was created by an enormous volcanic explosion over 750,000 years ago. Its sedimentary rocks contain magnetic reversals, evidence of historic switches in the Earth's magnetic poles.**

So by analysing the angle of the magnetic minerals in sedimentary rocks, geologists can determine roughly how far away from magnetic north (or south) the sediments were when the rock was formed. From a geographic perspective, geologists can work out the latitude (north–south) orientation of a sedimentary rock when it formed but not the longitude (east–west).

Using this methodology, Dr Harland analysed ancient dropstones found in sediments that he collected from two large islands in the North Atlantic Ocean: Greenland and Svalbard (which belongs to Norway). Harland had spent more than 40 years mapping the Spitsbergen islands, so he was well acquainted with these northern outposts with their dramatic ice formations, snow banks and polar bears. Yet Dr Harland's palaeomagnetic analysis showed that these islands had travelled far. In 1964, he revealed the first palaeomagnetic data proving that, between 720 million and 660 million years ago, glaciers deposited dropstones in tropical areas. He argued that these findings indicated that the Earth had once experienced an ice age of incredible proportions, one that covered the planet in glaciers and even brought ice to the tropics.

In the years since Dr Harland's seminal paper, further palaeomagnetic analysis of sediments with dropstones from around the world showed that these findings were not isolated. Dozens of papers documenting the presence of glacially deposited dropstones near the Equator have appeared, suggesting that Dr Harland was right: there probably had been glaciers in the tropics.

One of the dropstones embedded in the coastline of eastern Canada.

There are some geologists who question the palaeomagnetic data but, for the most part, the idea of there having been glaciers in the tropics between 720 million and 660 million years ago is widely accepted. What is not agreed upon is the extent of the freezing. Nobody is really sure whether Snowball Earth simply covered the continents in ice or if it froze the oceans as well.

Climate modellers have pored over calculations on how much heat from the sun is absorbed by land, water and ice. In the 1960s, Mikhail Budyko, a well-known Russian climate modeller, calculated that if the Earth had been covered by as many glaciers as the dropstones suggested it had been, the light-reflecting ice and snow on the continents would have reflected large amounts of heat from the sun and sent it back out into space. This reflection of light and heat, which can give skiers sunburn at below-freezing air temperatures, is known as the albedo effect. It would have led to the planet getting colder and colder. Budyko theorized that such cold temperatures would have caused the oceans to freeze entirely, and that would have increased the albedo effect further because ocean water, which is dark and readily absorbs warmth from the sun, would have become reflective with ice and snow cover and made temperatures on the planet plummet.

When first undertaken, this number crunching was useful because it explained how a Snowball Earth scenario could have come to pass, but Budyko argued that it would have been impossible. His modelling showed that if the oceans had begun to freeze, the heat reflection created by snow and ice would never have stopped – the Earth would have been locked in the freezer. The fact that today the Earth is not frozen was enough for Budyko to conclude that such a severe ice age never took place.

Dr Budyko's findings ignited fiery debate, and many geologists agreed with him. They found it hard to believe that the oceans, even when exposed to extreme cold, could freeze solid since they are so deep, have so many currents and, most importantly, hold so much heat (water is a powerful absorber of heat). Added to this were findings of sediment that appear to have been deposited by glacially fed streams during the worst of the Snowball Earth scenario. If the planet had frozen over entirely, how could there have been streams coming from glaciers? There should not have been any opportunity for glaciers to melt since temperatures were supposedly extremely cold.

Dr Joseph Kirschvink, one of the geobiologists who has researched Snowball Earth.

The disagreements and conflicting evidence have led some geologists to propose that perhaps the oceans froze only slightly, and that there were some periods of time when the planet was warm enough for some water to melt from the glaciers. Geologists like to call this alternative theory Slushball Earth. In this scenario, the glaciers came and went, creating more of a slushy mess than a completely frozen surface.

The debates about how solidly frozen Earth actually was continue today, but few dispute that the planet became extremely cold, and that most of it would have been covered in ice. However, this still left the problematic runaway albedo effect that Dr Budyko first identified.

For more than 30 years, geologists could not work out how the planet escaped from such frigid circumstances, where all of the sun's heat was reflected away by ice. But then came a proposal from Joseph Kirschvink, a professor of geobiology at the California Institute of Technology, and another polar explorer. He grabbed a lot of attention because he theorized that carbon dioxide was the planet's saviour and allowed it to escape from the cold grip of Snowball Earth.

It is ironic that, today, carbon dioxide gets all the bad press. It is seen as a troublesome gas produced by combustion, which contributes to the greenhouse effect that threatens our planet with dangerous increases in temperature. The reason it does this is because it absorbs heat from solar radiation, making the Earth warmer as a result.

Under normal circumstances on the planet, assuming there were no Snowball Earth and no humans to create large-scale pollution, carbon dioxide is released in large amounts by volcanic eruptions and absorbed when water reacts chemically with certain types of rock to form sediments. This release and capture of carbon keeps the amount of carbon on the surface of the Earth in balance. As volcanic eruptions pump more and more carbon dioxide into the atmosphere, the planet becomes warmer, but this is where the balancing cycle begins. Temperatures rise so water in the oceans evaporates more easily. With more water in the air, storms become more common, and these increase the amount of water that strikes surface rocks and causes erosion-based chemical reactions that remove carbon dioxide from the environment.

Interestingly, one of the great fears associated with global warming is that storms will become bigger over time as warmer temperatures bring more water into the air. While these bigger storms are highly destructive to human society, they are, in effect, the planet's way of increasing erosion and dealing with all of the extra carbon dioxide that humanity is flinging into the atmosphere.

—

*The debates about how solidly frozen Earth actually was around 700 million years ago continue today, but few dispute that the planet became extremely cold, and that most of it would have been covered in ice.*

Dr Kirschvink proposed that in a Snowball Earth scenario, with ice covering the continents, erosion and the interaction of water with rock would have come to an abrupt halt. Any water falling on the continents at that time would have fallen as snow or frozen into ice instantly upon hitting the ground. The planet's mechanism for controlling carbon dioxide levels through erosion would have shut down entirely.

Nevertheless, while carbon dioxide control on the planet came to a halt, the thick layers of snow and ice would have done nothing to stop volcanoes from erupting. Even under its skin of ice and snow, and with most heat from the sun being bounced back into space, the Earth's core was still searingly hot. Sporadic volcanic activity would have continued to melt through the icy surface of the planet and blast out lava. Each eruption would release vast quantities of volcanic gases into the atmosphere, and with no erosion to capture that carbon dioxide and trap it back in sediment, carbon dioxide is likely to have reached tremendous concentrations in the atmosphere over the course of millions of years.

The precise levels of carbon dioxide reached are not known, but computer models suggest that, in order for a completely frozen Earth to be thawed by a greenhouse effect created by the gas from volcanoes, carbon dioxide levels would have to have reached roughly 350 times their concentration on the planet today. This would have taken a

Research in the Canadian Rockies is contributing to the continuing scientific debate surrounding Snowball Earth.

long time, some tens of millions of years but, eventually, with such suffocating levels of carbon dioxide, temperatures would have warmed enough to melt sea ice and expose ocean water. And the great thaw would begin.

Water, because it absorbs heat so effectively, would have accelerated the melt by collecting even more of the sun's energy. The water would also have started to evaporate in the extreme warmth. This, in turn, would have further enhanced the warming effect, since water vapour, like carbon dioxide, is a greenhouse gas. Indeed, once the Snowball Earth scenario started coming to a close, the end would have arrived quickly. Within a short period in geological terms, there would again be running water on the planet as it poured off the great ice sheets that had gripped the Earth for so long.

———

The Snowball Earth theory provided researchers with an answer to how glacial dropstones might have been deposited in the tropics. They now had an explanation for how the planet managed to release itself from the self-perpetuating freeze. But when they looked again at their geological specimens, other rocks did not make sense. There, right on top of the glacial dropstones, were carbonates – minerals that form most commonly in very warm, shallow seas.

Dr Kirschvink had an explanation for how carbonates fitted into the story. If carbon dioxide reached 350 times its modern levels, this would have caused the planet to become hot and would have led to warm oceans and the formation of carbonates in them. In the searing atmosphere of the post-Snowball Earth era, warm oceans and shallow seas would have been common.

Support for Budyko's albedo theory and Kirschvink's carbon dioxide theory came from geologist Paul Hoffman at Harvard University. He found his evidence for Snowball Earth and the great thaw that followed in the soils of Africa.

In the rock record of Namibia, the team found dropstones followed by carbonates, followed by more dropstones, followed by more carbonates. Hoffman and his team explained in the journal *Science* in 1998 that these strange patterns suggest that there was not just one big ice age followed by a very warm period but, instead, multiple ice ages with multiple warm periods in between.

Biology lends some weight to this idea of climatic oscillations. Aside from rainwater interacting chemically with rocks to absorb carbon dioxide, if the Earth had a warm environment, the seas would be rich from nutrients being washed off continents by large storms, and photosynthesizing single-celled organisms would have experienced population booms all around the planet. These organisms, like plants, inhale carbon dioxide and exhale oxygen. If their numbers were great enough, they could have dramatically increased the carbon dioxide control effect produced by water-rock interactions. Indeed, while nobody is sure what caused the extensive glaciations, it is theorized that photosynthesizing organisms played a key role in removing large amounts of carbon dioxide from the atmosphere. It was this significant reduction that sent the world into the freezer in the first place. Regardless of the exact causes, which will probably never be known, it seems the Earth jumped from the freezer into a furnace, and back again, many times.

———

On the Columbia Icefield, Hazel Barton tries to keep thoughts of a warm Earth away. She has now reached her destination: the tip of the glacier, a wall of ice sloping up beyond her head, with wide, blue cracks receding into blackness. The dripping is constant, and streams of water flow from the bottom of the ice, forming a wide, foaming river a little further down the slope.

It is in this mud, rock and 'dirty' ice that there is life in abundance - microscopic cells that are built for sub-zero survival. While most people think of glaciers as sterile places, Barton sees them as veritable rainforests that can tell us a lot about how life made it through the Snowball Earth era.

Where exactly life held on is hard to say - life was still single-celled, and fossils from the time are rare to find - but biologists like Barton, who study modern single-celled organisms, think that there were a few places where life went during this difficult period. Some suggest that life survived near thermal vents like the smokers at the bottom of the ocean. Even if the oceans were entirely frozen over, these regions would have supported life as the planet went through its freeze-thaw seesaw experience. However, it is unlikely that these would have been the only refuges for life.

Barton studies single-celled organisms called cyanobacteria, which generate nutrients for themselves using the sun's energy while surviving in seemingly inhospitable conditions. If they were to dry out, many of these organisms would become dormant, shutting down for years without the need for any nutrients. But there is mounting evidence that such bacteria can thrive in what we would think of as the most inhospitable conditions. 'Microorganisms that live in these harsh environments are called extremophiles,' Barton explains. 'They have this amazing ability to adapt to the harshest environments, to close themselves down. You can take all the moisture out of them, you can freeze them, you can bury them a mile down in ice, and you can stick them there for hundreds of years and then, when conditions become more favourable, they can be resurrected and continue living like nothing happened.'

For example, many of these organisms can manage to get water from the ice itself in arctic environments. At the base of glaciers, where

Hazel Barton examining extremophiles on the Columbia Icefield.

lots of sediment collects in high concentrations, the cyanobacteria can obtain many of the mineral nutrients that they need to survive. With food and water (and special anti-freeze chemicals to stop them from turning into microscopic ice cubes), the bacteria have all they need. So when Barton peers carefully down a microscope at what look like brown smears of dirt on glaciers, thriving ecosystems come into view. The smears of dirt are groups of single-celled organisms clinging to bits of rock on ice in a place that would be utterly hostile to most life.

For Barton, what makes cyanobacteria so important is that they are ancient. She argues that they are living fossils, showing what life was doing during the days of Snowball Earth. Given that these organisms can survive today under such severe conditions, she is left in little doubt that they survived extreme conditions long ago.

'You had a skin of microbes on the surface of the planet, and you had these organisms living in the thin gaps where the glaciers contacted the rock,' Barton says. 'And those few organisms hanging on in there over millions of years were numerous enough so that when it became more favourable, everything was able to take off again.'

Remarkably, there is extensive life inside and underneath the ice of glaciers. Looking a bit like bacteria, the single-celled organisms living in these environments are from a family of organisms called archaea. These unusual cells, which are actually more closely related to animals, including humans, than unicellular bacteria, have an amazing ability to obtain their energy from the chemistry of rocks. They literally consume materials such as iron to survive.

———

But life did more than merely survive during the Snowball Earth years. Life came out of the freezer–furnace experience dramatically more

A scanning electron microscope picture of *Methanosarcina mazei* archaea, an extremophile. Archaea are single-celled organisms that are like bacteria, but also have characteristics of other organisms. Clumps of these cells are seen here – they can survive in conditions that would be deadly to most other forms of life.

complex than it had been before. Just after the end of Snowball Earth, we find organisms constructed of more than just single cells. Initially, this seems odd. Why, after more than 2 billion years of existence on the planet, and an extended period of extreme conditions, did life suddenly develop complexity?

The answer that Hazel Barton and many palaeontologists propose is that resources – energy, nutrients and water – abruptly declined during the Snowball years. As the planet froze, surface oceans transformed into ice, and most single-celled organisms died out because there was no longer any way for populations to obtain sunlight and soak up nutrients from the water. This mass extinction also struck predatory organism populations. Predators that fed on sunlight-absorbers starved to death as their food source died out; in turn, the predatory cells that fed on these predators began to starve as well.

Under such circumstances of mass starvation, any single-celled organisms with characteristics that favoured their survival – such as the ability to live near boiling thermal vents, to live on ice instead of water, or to live off chemicals in rocks – would have bred and created small populations in localized environments. These isolated refuges, which were functionally a lot like islands in vast oceans of ice, were probably the keys to the burst of evolution that was soon to follow.

———

Islands have a dramatic effect on evolution. When organisms reach islands, they usually arrive as just a small population and breed among themselves, but they are also exposed to new and unfamiliar pressures and opportunities. These effects cause founding populations on islands to quickly become different from the original populations; characteristics that are rare in mainland populations but present in the founders of the new population can become concentrated and common in the founder population.

Over time, this concentration of what was once a rare trait can make the island population so different that it becomes a different species entirely. Consider the following hypothetical example.

Suppose there is a population of seed-feeding parrots on a continent. Most are green but a small number are blue. A storm hits the continent and two unfortunate parrots, a blue male and a blue female, are blown out to an island that recently formed off the coast following a volcanic eruption. As these two parrots breed, they will create offspring that are blue because both of the parents are blue. As these offspring breed with one another – as there is nobody but family to breed with – they are also going to create blue offspring.

These circumstances would lead to an eventual parrot population that carries entirely blue parrots, rather than the mix of green and blue parrots on the continent. Does this alone create a new species? No, because if one of these blue parrots were introduced to a parrot

from the mainland, the mainlander would still view the island parrot as a potential breeding partner since blue parrots are sometimes encountered on the mainland.

For a new species to form, other changes must occur. Like the mackerel in Chapter 2 that varied by weight, size and speed, the parrots would also have characteristics that vary in the population as a result of the diversity created by sexual reproduction. Some mainland parrots might have slightly bigger beaks than others; some might have sharper claws; and some might have longer necks. The variations are much like those seen in human populations, with some people being taller, some having blond hair, and some being more muscular. Among the mainland parrots and among humans, these variations are noticeable, but not extreme. However, for a couple of parrots blown out to an island by a storm, these variations can quickly create tremendous diversity.

Let us start with beak size. If the founding female of the island parrot population had a long beak, just a millimetre or two larger than average, and the founding male had a small beak, just a millimetre or two smaller than average, this would allow their offspring to show a range of beak sizes, depending upon which parent's genetic information they inherited.

This variation in beak size would exist on the island as each generation of blue parrots dominated the landscape. However, at some point the island parrots would encounter trouble because there would not be enough seed to support their growing numbers. Starvation would follow, but any parrot on the island that found a way to get nutrients from a source other than seeds could live longer and breed more than its kin.

Perhaps there are nuts on the south side of the island, and here the parrots with a particularly large beak manage to crack open the nuts and feed. Therefore, the long-beaked parrots on the south of the island would live a long life and generate lots of offspring that all carry the long-beak genetic information. Since nuts are found only in the south, these big-beaked, nut-loving parrots would probably start breeding with one another instead of with smaller-beaked parrots in other regions of the island. This localized breeding (behavioural isolation) would further lead to the concentration of long-beaked parrot genes in the south and create a population of nut-feeding specialist parrots.

Meanwhile, seeds have now become scarce on the island, but some birds - those with sharper claws - discover that they can catch insects and eat those instead. Parrots with slightly longer necks also adapt to reach food resources, like fish, that other parrots on the island cannot. After years of pressure from limited seed resources, several varieties of parrot would eventually evolve: nut-eaters to the south, fish-eaters around the coast, and insect-catchers that prefer the swamps and marshes where mosquitoes breed.

The variety of animals on different islands – such as parrots in the Caribbean – is a classic example of evolution in action.

# Ways of Looking at Fossils

"I never tire of the process of learning. Acquiring information and understanding is hugely rewarding and great fun. Working closely with palaeontologists at important fossil sites around the world has given me some insight into their work. I have learnt a great deal about how ancient creatures looked and moved, and I've come to understand what to look for on a fossil expedition, and, just as importantly, how to look for it.

I found fossil excavations extraordinarily revealing of the way in which palaeontologists work these days. When I first became interested in fossils, the process of understanding your findings was pretty simple. Having collected a few fossils you then did some research to find out the names of the organisms that made them. If they didn't have names, you came up with a name, without ever being so arrogant as to name it after yourself.

Now there is no point in simply picking up a fossil and looking at it. You have to know how it was positioned when it was found, in which direction it was pointing, how far it was from water or rocks and what the layer of sediment around it contained. This information allows you to paint a detailed picture not just of the organism itself, but of its whole ecosystem; then we can deduce how different organisms interacted with each other and with their environment.

This way of thinking has broadened our knowledge about early life dramatically, and it's thanks to the sharp minds of palaeontologists that we have been able to learn so much from the fossil record."

**This trilobite fossil (left) was found in the Burgess Shale in Canada. Visiting fossil excavation sites such as Mistaken Point (right) is revealing of the way in which palaeontologists like Guy Narbonne work.**

Over time, the island parrots would become so different from mainland parrots (both genetically and physically) that birds from the two populations, should they meet, would no longer be able to breed. Indeed, it is likely, with the island parrots becoming so specialized, that even the various specialized populations would not be able to breed with one another. The island isolation that began when those two unfortunate blue parrots were blown from the mainland so long ago has eventually created several new species.

Our parrot scenario is a simplified example of how a new species is created, but a sequence of events very similar to this took place with finches on the Galapagos Islands. Millions of years ago, a small population of finches somehow ended up on the islands. But since then they have evolved into a wide variety of remarkable forms.

While birds and islands are the easiest way to illustrate how isolation fuels evolutionary change, neither birds nor islands are essential for new species to evolve as a result of isolation. Mammals isolated by rising mountain ranges, insect populations cut off by newly formed rivers, and reptiles trapped at oases in the middle of vast deserts have all experienced the effects of isolation fuelling evolution, too. With this in mind, let us return to Snowball Earth.

These finches from the Galapagos Islands gave Charles Darwin important insights into how speciation takes place when a population becomes genetically isolated.

If an extremely harsh ice age led single-celled organisms to be trapped in small, isolated populations divided by large ice-filled landscapes, the same evolutionary mechanisms that caused our imaginary parrots to speciate would also have caused the ice-bound single-celled organisms to do so. Then, when the planet warmed up and melted away all the ice, the newly evolved organisms that arose in those 'island' populations would spread, mingle and compete before being isolated again by another freeze cycle. The mixing would spread variety far and wide and provide the genetic fuel for further evolution as the ice encroached once more and re-established population isolation.

In effect, palaeontologists propose that each Snowball Earth episode created numerous island effects that fostered many behavioural isolation events, and these events drove the evolution of single-celled organisms towards the multicellular life forms of today.

There is no question that Snowball Earth is an enticing theory. It is enough to motivate scientists like Hazel Barton to haul bags of specimen jars and heavy brass microscopes up icy mountainsides. The evidence of glacial dropstones in the tropics is so compelling and the evolutionary explosion in diversity following the era of hot and cold fluctuations is so similar to what we have seen happen on islands that it has captured the imagination of both scientists and the wider public. Unfortunately, some fossil evidence being dug up by Susannah Porter, a palaeontologist at the University of California, Santa Barbara, and a team of colleagues suggests that things were not necessarily so simple.

Porter is a particular fossil hunter. Fossils are old, she knows, but there is old and then there is really old. But thanks to the active nature of the Earth's crust and the destructive power of the ice age, pre-Snowball rocks are hard to find, and fossils of the squishy, single-celled organisms that were around before the big freeze are even rarer. But Porter has been fortunate to work on something quite extraordinary: fossil beds containing single-celled organisms from 15 million years before glaciers started appearing in the tropics. She believes these fossils tell an amazing story, that something other than the freeze-furnace fluctuations of Snowball Earth initiated evolutionary change.

Dr Porter and her colleagues have discovered fine-grained rocks dating to before Snowball Earth at the bottom of the Grand Canyon. These specimens contain fossils that show a major ecological shift in the world of single-celled organisms. In older layers, there are single-celled species that would have been generating their food from sunlight, but they become increasingly rare in younger rock layers. Eventually, they are replaced by single-celled species that are distinctly different.

This discovery represents a potential wrinkle in the Snowball Earth theory. If ice had been covering the planet, the differences between

# Adventures in Palaeontology

Ecologists often find themselves travelling deep into jungles to study biodiversity, facing the threat of disease, parasites and venomous snakes along the way. Marine biologists must deal with sharks, dangerous currents and storms that can sink their boats. Volcanologists often venture within spitting distance of the fiery features that they study to collect their data. By comparison, palaeontologists seem to have it easy; all they have to do is travel into typically arid terrain, in shirtsleeves and sunhats, and do a bit of digging. This is not the case, though, for researchers studying fossils that formed just before the Snowball Earth event took place.

Palaeontologist Susannah Porter and a team of colleagues became aware of layers of rocks made from fine sediment called the Chuar Group that contained fossils from just before glaciers appeared in the tropics. They wanted to study these layers in depth to get an understanding of what was happening to life just before the Snowball Earth event occurred. Getting to the rock layers was an adventure.

The Chuar Group's only exposure on the surface of the Earth is in a remote location at the bottom of the Grand Canyon. Extreme weight restrictions meant that helicopters could ferry only the researchers in and out of the canyon, not their heavy rock samples. For Porter and her colleagues to get to the Chuar Group and collect samples, they either had to face the fierce heat of the Grand Canyon trails or brave the region's legendary white-water rapids. During their many collecting visits, they would experience both.

On the river, the team had to navigate their rafts 150 miles through the canyons to a location where they could ultimately join a track. At one point, their 6.7-m (22-ft) raft flipped in the white water. While their precious samples remained safely stowed, the team and their river guide were thrown from their craft into the raging water, the research team's kitchen equipment sank, the guide lost his glasses, and the whole team were forced to swim for their lives through the dangerous rapids.

When the team attempted to set out on foot, things were just as treacherous. The 17-mile trail was not well maintained, and they had badly miscalculated the amount of water that they would need. Miles before they reached the rim of the canyon, they were suffering from extreme dehydration. Desperate for water, they fortunately found some in a spring and immediately sat down to drink. It was not until later, when they started to rehydrate, that they realized the spring was surrounded by poisonous plants. Once again, a potentially fatal accident was narrowly missed.

Working at the site itself was dangerous. Rattlesnakes, mountain lions and scorpions are all residents of the Grand Canyon. A simple bite or sting can quickly turn from being a minor medical problem to something far more severe, difficult to treat and potentially fatal.

The threats are many, but the fossil finds for palaeontologists like Dr Porter more than make up for them. Many palaeontologists seem to have inherited the adventurer's spirit from the pioneering fossil hunters of the past. They are fond of visiting vast untamed wilderness – harsh mountainsides, polar icecaps and arid deserts – and having to travel to such remote locations is part of the attraction. Collectors of these natural relics are explorers and adventurers. And every fossil has a unique, multi-layered story to tell, one that spans many millions of years from its amazing creation to its recovery.

The rock layers of Mistaken Point were once mud lying at the bottom of an ocean.

these organisms would make perfect sense, since ice cover would make it difficult for ocean-surface-dwelling, sunlight-collecting, single-celled organisms to survive. However, glaciers did not appear in the tropics until 16 million years later, long after these fossils were formed. While Dr Porter's findings do not disprove the idea that Snowball Earth played some role in creating a series of stressful events that accelerated evolution, they also suggest that Snowball Earth was not uniquely responsible. Why did the plankton disappear? It looks as if something stressed life long before ice even started spreading.

What that stressor was is difficult to say. Porter's team studying the fossils in the Grand Canyon found evidence in the rocks of iron concentrations increasing in ocean water. As we saw in Chapter 2, the formation of banded iron occurs when oxygen is readily present in the atmosphere – iron rusts out of water and then falls to the ocean floor. If iron concentrations in water were rising, this suggests that oxygen was becoming rarer in the air, hinting that something on the planet was changing the balance of gases in the atmosphere. What that was exactly we do not know, but these findings are not alone in raising questions.

There is a fossil site in California's Death Valley that dates to the middle of the Snowball Earth era, a period that was extremely cold. But this particular fossil sample shows a mix of sunlight-feeding and predatory single-celled organisms. How could such a diversity of life be present in the midst of a time when conditions were so harsh?

Again, we don't have all the answers, and fossil finds like these merely raise more questions about just what sort of evolutionary effects Snowball Earth actually had on those early organisms. But the adventure continues. Scientists like Hazel Barton and Susannah Porter will endure their seemingly mad expeditions to the harshest, wildest and remotest parts of the planet. They will brave freak storms, cross ferocious rivers and face unexpected dangers. The fossils will continue to be collected, and the rocks will continue to be analysed. And perhaps, one day, they may conclusively prove – or even disprove – the theory that the harsh conditions of Snowball Earth drove the evolution of more complex life.

**Rocks dating to before Snowball Earth have been found at the bottom of the Grand Canyon, Arizona.**

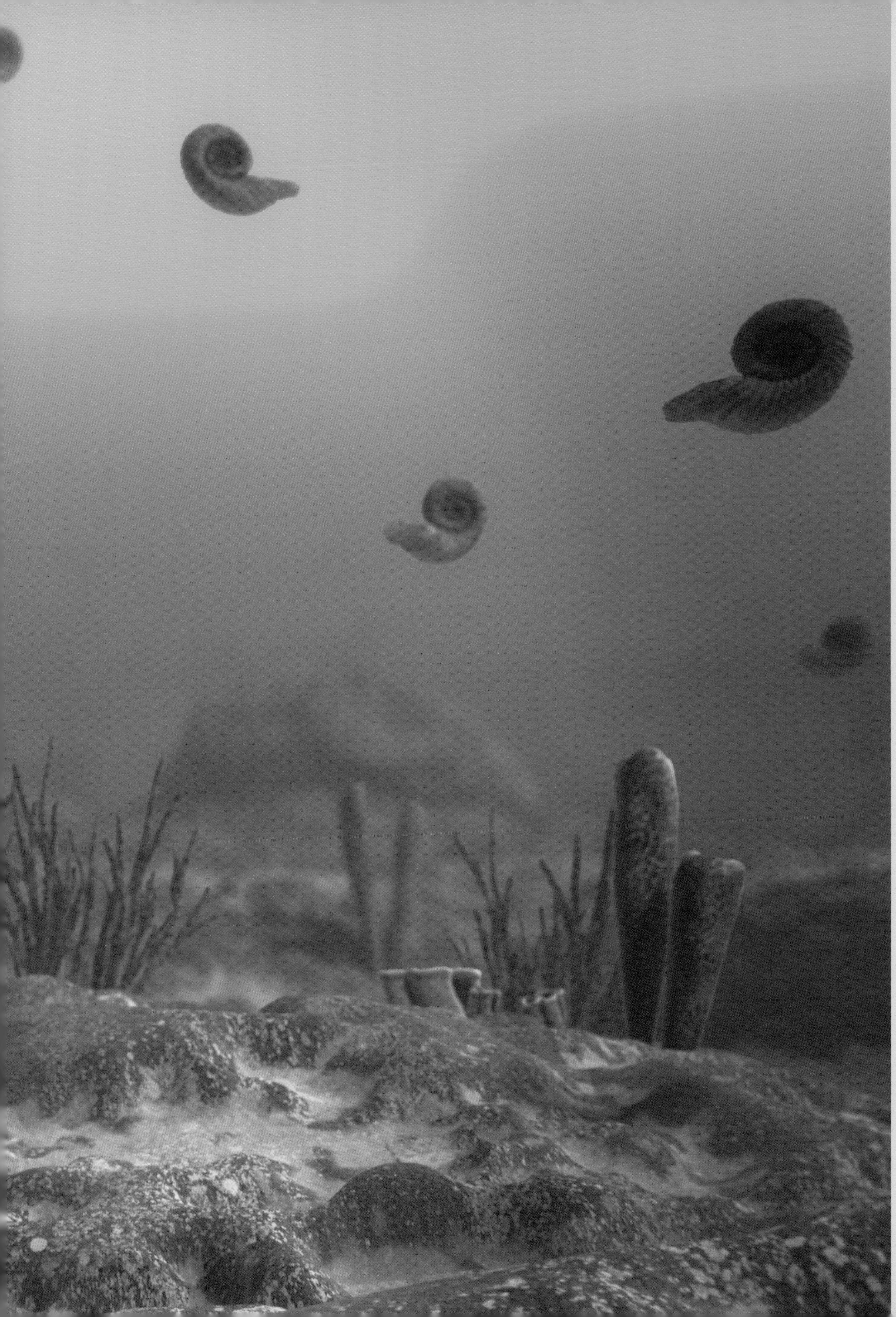

CHAPTER 4 · THE DAWN OF ANIMALS

THE SEAS THAW, the glaciers retreat, the icecaps shrink. Snow turns to rain, the air warms and that bright, dazzlingly white Snowball Earth returns to its original blue colour. Once again, there is flowing water all over the planet, with waves crashing against cliffs, streams trickling through valleys, and waterfalls pounding into deep plunge pools far below. Water pours from the continents in torrents, and floods the oceans with millions of years' worth of ground-up rock and minerals.

What a feast! It would have been irresistible for the cyanobacteria and the archaea. Trapped for so long in their ice islands, eking out a meagre existence from the nutrients scoured free by the ice, these cells would have had a feeding frenzy following the end of Snowball Earth.

The global glaciation of Snowball Earth has churned up the planet and fertilized the earth. An event that nearly wiped out life has instead put masses of ground-up rock into the oceans, and that rock has acted like inorganic fertilizer for the protoplants, which are about to enjoy a huge growth spurt.

As they invaded the newly ice-free world, the cyanobacteria experienced population booms that released vast quantities of carbon dioxide into the atmosphere while consuming oxygen. However, their consumption of oxygen and release of carbon dioxide did not, for some reason, send the Earth back into a superheated greenhouse effect scenario. At this time, around 650 million years ago, Snowball Earth finally came to an end.

The global glaciation of Snowball Earth came to an end around 650 million years ago.

Palaeontologists suspect that the planet was released from its freeze–thaw cycle because this time there was something new in the mix, something converting the carbon dioxide back into oxygen.

As the glaciers retreated and the runoff from them filled the sea, the runoff fertilized all of the ocean-dwelling bacteria that used sunlight, and made early forms of plants in the oceans grow. The huge colonies of protoplants began photosynthesizing more furiously. Since oxygen is a by-product of photosynthesis, these plants pushed large amounts of it into the atmosphere and into the oceans. Consequently, the planet experienced a dramatic rise in oxygen levels.

# Return to Heron Island

" I visit Australia most years, and have been to Heron Island, on the Great Barrier Reef, several times over the last 50 or so years that I've been working in television. When I first came here we had to stay in rather basic little cabins, but since people have started travelling more and more they've really developed the island. It's a wonderful place for people to come and see the wildlife, and it's also really significant in terms of ecology and research.

Heron Island is a cay, composed of coral, right on the reef itself, which means you can easily get out to film corals and other sea life. Because the reefs around here are so shallow, when you're walking around in the water, it's easy to imagine what it must have been like in those Late Ediacaran seas when corals and sponges were the most complex things around.

Marine researchers can bring back specimens to the lab on the island and then release them quickly into the ocean after they've finished their studies. This is really important because animals like corals are tremendously delicate, and don't really like being pulled out

of their environment. For research purposes, it's better to transport corals as quickly as possible and get them settled in the lab before they become damaged.

On Heron Island, there's been some important research performed on sponges and on corals. Not many people realize how important these creatures are. Coral reefs act as a nursery ground for a great number of oceanic species of fish, including fish that provide humans with food.

If you lose the reef you lose this essential nursery, and the knock-on effects are profound. It is a tragedy that acidification of the ocean, caused by rising levels of carbon dioxide, is now established and rising fast. If the oceans get too acidic then corals can't secrete their calcium carbonate skeletons; they dissolve as they are produced.

The world's coral reefs are in a fairly bad state compared with how they were 100 years ago, and it's all because of human activity. If we stop polluting right now there is hope that the reefs will recover, but it's unlikely to happen in my lifetime."

**At the Heron Island Research Station on the Great Barrier Reef, scientists are working to understand how multicelled organisms first appeared on Earth.**

As we discovered in Chapter 2, even in the earliest days of life on Earth, ancient ancestors of animal cells are thought to have already evolved into predatory cells. These are believed to have consumed or collected smaller organisms, which eventually evolved into the mitochondria that help animal cells generate energy today.

Mitochondria – the extraordinary 'cells within a cell' that you find in every living animal cell (except red blood cells) on the planet today – are big consumers of oxygen; the more oxygen they can get, the better they function. So, if the early organisms that were beginning to flourish in the post-Snowball Earth age had lots of mitochondria inside them, it is possible that these microscopic power plants tipped the balance. They consumed the oxygen released by the photosynthesizing cyanobacteria and archaea, and pumped tonnes of carbon dioxide back into the atmosphere. This release of greenhouse gas kept the Earth warm and brought an end to the relentless hot and cold cycles of the Snowball Earth period.

*What most people don't realize is that ocean sponges are actually animals with a unique biology that holds the answers to some very important questions about the story of life on this planet.*

This theory is supported, in part, by the fossil record. We find that all the organisms dating back to those balmy days when the world was escaping from its freeze–thaw trap had numerous cells in their 'bodies'. The many cells that are present in animal bodies are stuck together by materials that demand high concentrations of oxygen, so these animals could have lived only if there were high levels of oxygen in the air. These organisms would have released carbon dioxide as they fed, grew and reproduced.

With this extra oxygen, other microbe-like early animals had the ability, for the first time, to produce a substance called collagen. With the advent of collagen, suddenly these early animal cells had a glue that could stick them to other cells. So just how important was the development of collagen? The answer is that it was critical. Today, within every animal, including humans, cells are stuck together with collagen.

One animal that survives to this day in marine habitats offers clues to an extraordinary evolutionary event. It is one of the most primitive animals we know of, but its basic body structure has enabled it to survive nearly 600 million years of evolution. It is the sponge.

Sponges are collections of simple cells that have come together and have stuck to one another to create a single larger organism. They have neither a digestive system, a nervous system nor a blood circulatory system, and they get their food and oxygen simply by pumping sea water through channels in their body. But they give us an indication of how cells first clumped together to form bodies comprised of multiple cells.

This was a landmark step on the evolutionary road that led to the modern, complex animals alive today.

At first sight, ocean sponges seem like truly alien creatures, their form barely resembling any animals familiar today. Until the 1950s, ocean sponges were pulled from the sea in vast quantities and sold around the world for use in bathrooms. Indeed, sponge collecting was big business until demand exceeded availability. As is so often the way when humans find a use for a natural product, sponge populations in many regions were almost wiped out. Of course, most bathroom sponges today are manufactured artificially, but what most people don't realize is that ocean sponges are actually animals with a unique biology that holds the answers to some very important questions about the story of life on this planet.

———

The sponge, whose primitive forms first evolved in the post-Snowball Earth era when oceans were rich in nutrients and oxygen, has attracted considerable interest from scientists studying this period in the evolution of life. Sponge cells, like human cells, generate a specialized structural protein called collagen. And some researchers, like sponge biologist Dr Bernard Degnan at the University of Queensland in Australia, think that it was the evolution of collagen and allied materials that was critical to the evolution of animals.

On some beaches in Queensland, you can scan the tide line or roll up your trousers and wade into the water, then feel around and soon you'll brush against a squidgy mass, a piece of sponge. Of course, Dr Degnan is a little more systematic when he collects sponge specimens. But his study of these innocuous-looking organisms and their collagen has had a profound impact on our understanding of how multicelled animals probably evolved.

———

Collagen proteins – along with a raft of other molecules found outside modern sponge and human cells alike – are instrumental in allowing cells to link together in an organized structure. Collagen is produced by individual cells and released to the outside environment. Other cells can then attach to the collagen framework and release collagen structures of their own. In a sense, collagen is the biological glue that holds together the cellular world, explains Dr Degnan. It is a framework created by cells for cells to live on.

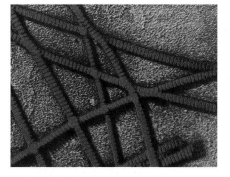

Collagen is the most common protein in our body and forms the framework of our skin.

# Sponge Embryos

66 The exact moment when a 550-million-year-old cell began to divide was captured in a 3D image in South China in 2006. The astonishing discovery of a group of fossilized embryos shed light on the early evolution of complex life, but what are they?

Dr Philip Donoghue believes they are the fossilized remains of sponge embryos, and that they follow the same pattern of cell division that is seen in humans. Using a gigantic microscope in Switzerland called the Synchrotron, he looked inside the fossils. The Synchrotron is the only X-ray-type machine that can provide the kinds of resolution necessary to see all of the fossilized embryos. Powerful generators fire high-energy

electrons around a circular tube at close to the speed of light. As they round the bend, the electrons emit X-rays so powerful they can penetrate solid rock – or the tiny fossils.

'It was astonishing,' Phil told me. 'It was a real eureka moment that you could get to the very finest levels of fossilization, the very finest detail that the fossil record could ever give up using this kind of technology.'

Phil used data from the Synchrotron to build the 3D picture of the fossils and draw his conclusion. 'We know it's a fossil embryo because it's surrounded by a preserved egg sac,' Philip said. 'Using tomography we can see inside to the developing animal.'"

**The Synchrotron: a gigantic microscope housed in a building the size of a stadium that allows scientists to perform CAT scans on fossils.**

'We look for commonalities, for things that bind all animals together. So what does a human share with a chimpanzee, a tiger and a sponge?' Degnan contemplates. 'If we can find any common threads between humans and sponges, we're getting to the heart of the matter of multicellularity in the animal kingdom.'

Degnan, along with many other biologists, believes that collagen must have been around in order for single-celled animals to evolve into multicellular ones. Without it, there would have been nothing for them to cling to. Moreover, in order for collagen to be manufactured by cells, oxygen needs to be in ready supply. So, with levels of oxygen increasing in the post-Snowball Earth world, collagen could be readily built. All the conditions were perfect for our ancestral cells to become sticky-tape factories.

The evidence supporting Dr Degnan's assertions is strong. All animals alive today – including humans – have collagen in their cells. While it is possible that collagen appeared independently many times after the rise of animals, the chances that this was the case are exceedingly small. It is much more likely that collagen appeared, by random chance, in just one early organism. It turned out to be useful; the cell stayed alive, reproduced and passed the genetic code for this new sticky protein to its descendants. And so collagen, which proved to be advantageous for life, was passed through the generations.

Genetic analysis of modern single-celled organisms supports this argument. Some of these organisms produce materials that look a bit like collagen, but it is not the real stuff. These not-quite-collagen materials are manufactured by cells using instructions from DNA that, if tweaked a bit, would create collagen. In fact, these cells are on the verge of independently developing collagen for themselves – and, therefore, they are just 600 million years behind the rest of the animal world. Nevertheless, the fact that some single-celled organisms have almost independently evolved collagen and that collagen is present in all animals, whether they are as simple as worms or as complex as sea-turtles, makes a strong case that the substance played a key role in the evolution of the first animals.

But the presence of collagen alone did not lead to the rise of animals. For multicellular animals to appear from a single-celled existence, there had to have been some sort of advantage in clumping together – an evolutionary pressure that caused multiple single-celled organisms to come together and use the collagen they could create. Degnan argues that since sponges are the simplest animals alive today, it is in them that we can find the answer.

At first glance, sponges appear to be simply large collections of cells that, through the sticky nature of collagen, stay together. There's a classic experiment that shows how they may have first clumped together. First, a sponge is cut into small pieces. Next, it is pushed

Stages from a time-lapse series taken over seven days showing the capture and digestion of a mysid shrimp by the carnivorous sponge *Asbestopluma hypogeal.*

through a filter on the end of a syringe into a tank of water. The filter separates the sponge cells so that the tank fills with a cloud of diffuse, individual sponge cells. The animal has now been broken down into individual cells. This may seem like a brutal act committed against a living organism, but to a sponge it is nothing. It then goes on to do something quite astonishing. After a few weeks, that sticky collagen has bound the cells up, and there are small clumps of what can only be sponge forming in the bottom of the tank. 'What you'll find is that these cells will start to move towards each other and they'll start to link up again,' Degnan explains. 'Sponges have this amazing capacity to regenerate themselves. What we can do is actually rebuild a sponge from the cell level up. Three weeks into the experiment we have a miniature sponge.'

This same pattern of growth is seen in the embryos of all animals today and it gives us an insight into how the first animal bodies might have developed. 'From this classic experiment, we can infer a few things that happened 600 million years ago with the very first animals,' Degnan says. 'We can infer that there were cells coming together, they could adhere to each other, and that they used extracellular proteins like collagen to glue themselves together. They had the ability to communicate with each other and they had a certain amount of plasticity if you like, flexibility, that allowed them to interact and communicate with each other – and their tasks, their individual tasks as cells or groups of cells to give rise to something that's bigger and greater, a large macroscopic multicellular animal.'

Any cells that are genetically identical can recognize one other. All of a human's cells can recognize one other. They are covered in proteins that create a unique fingerprint that your body can recognize. Essentially, all you are is a colony of cells working together as a team, and that's what the sponge is trying to demonstrate. But a sponge is so primitive, you can separate it and its cells can still exist on their own. After being separated, those cells are constantly seeking each other to reform that colony, which is what those early cells would have done.

When the sponge cells are viewed under a microscope, you begin to appreciate how there is so much more to a sponge than meets the eye. Many of the cells in sponges have whip-like structures on them called flagella. These flagella move back and forth in the water, creating a current that draws water past specialized cells adapted to capture and collect nutrients that might otherwise shoot past. Then more cells help to transfer collected nutrients to the flagellum cells so that they can generate enough energy to continue whipping water and generating a current.

So that soggy, squidgy thing that sat on the side of almost every bath until the 1950s turns out to contain specialized cells that give it shape, structure and support. It is easy to picture its early ancestor: a blob

of flagellum-carrying cells, bound together by chance with collagen sticky-tape. The blob thrives because the tiny currents created by the incessant whipping of their tiny tails make them better at getting nutrients than the floating single cells, which can only wait for a chance encounter with their next meal.

The basic arrangement of modern sponges creates an irresistible evolutionary story. There are plenty of single-celled organisms with flagella today that are very efficient nutrient collectors. So if two or three of these organisms had come together during the hard times of Snowball Earth and benefited by simply being next to one another, it is easy to see that the arrangement would have ultimately proved beneficial, especially if harsh conditions caused food supplies to run short. Indeed, if a flagellum-carrying organism arose (via mutation or the variation created from sexual reproduction) that stayed partially connected to its parent, rather than floating off alone into the ocean, such an individual – and its many connected offspring – would have out-competed and out-bred nearby organisms by creating a stronger current with its flagella and feeding more often.

A group of mutant single-celled organisms that stayed connected after reproduction would have effectively been the first multicellular life on the planet. They were likely to have been the first animals.

How, though, do you get from a clump of flagellum-carrying cells, all stuck together but each competing for nutrients, to a structured sponge with several types of specialized cell? Again, biologists like Dr Degnan believe that evolutionary forces would have probably tightened the bond between cells that did not divide completely or break away from the family. Due to the natural variation in the characteristics of the cells, some of the flagellum carriers might have found themselves doing the bulk of the nutrient collecting while others were doing the bulk of the whipping. But as long as some nutrients were being shared across the partial connection that remained between the cells, natural selection would have favoured sharing if the cooperation between the cells increased overall survival.

So the whippers, who were getting most of their food via cellular connections rather than capturing food themselves, would have less need for machinery to capture food. Why run a redundant system? Why waste precious energy and resources on maintaining this system? Energy efficiency drives business transformation today, and it drove cellular transformation 600 million years ago. The flagellum-carrying cells discarded their redundant equipment to become lean whipping machines. They became more efficient, they could breed more often and, with time, they became more common.

In contrast, cells that were doing most of the food collection but not much whipping evolved in the other direction. Those cells that,

# Sponges not for the Bathroom

While most sponges passively collect nutrients from the water, there are a few today that live in very deep waters, often as far down as 5.5 miles. At these great depths, the nutrients that sponges normally rely upon in the open water are not readily available. To compensate for this lack of nutrition, evolution has driven these sponges towards lives of carnivory. They capture and kill small animals – a far cry from the passive life of nutrient filtering.

How exactly the sponges capture their prey is a matter of much debate. Those sponges that engage in this behaviour are both rare and not well studied because of the depths of their habitat. Even so, initial observations show that these animals use thin threads of tissue to snag small crustaceans. Once they have made a catch, the sponges appear to cover their prey with more and more thin threads. Ultimately, when they have fully covered and imprisoned their prey, the carnivorous sponges overwhelm and absorb the captured animal.

Most carnivorous sponges have, intriguingly, lost their typical sponge-like behaviour. They do not create water flows with their flagellum-carrying cells, and they tend not to have internal cavities for nutrient absorption. Clearly this branch of carnivorous sponges arose long ago when some sponges were deprived of food in the water column. Natural selection led one of these sponges – which, by some quirk of mutation, had developed a taste for meat – to survive and reproduce.

Today, carnivorous sponges appear to be rare but, in time, as exploration of the dark and mysterious fathoms below increases, it is likely that more of these fascinating, iconoclastic animals will be found.

Sponges take on many forms. They can be branching, blob-like or even crusts that grow over surfaces, but one of the most common forms has a cavity at the centre. These sponges have their flagellum-carrying cells positioned so that they draw water into the cavity where it is forced through sponge tissues. Inside these tissues, nutrient-collecting cells strain out oxygen, tiny organisms and other nutrients on which the sponge can feed.

The openings of this sponge cavity look a bit like non-moving mouths, and the similarity of the cavity interior to stomachs in other animals is striking. However, sponges are porous. They never developed sheets of cells that allowed them to create a distinctive inside and outside. It was the jellyfish, and their ancient kin, that evolved this characteristic.

**A large barrel sponge covered with crinoids, commonly called feather stars, in Papua New Guinea.**

through sexual variation, were produced with smaller, less fully developed flagella would have had an edge. They would have been more efficient and more likely to reproduce successfully because they were no longer pouring resources into developing a flagellum structure they didn't need. In effect, the partnership created by mutation would have driven the connected cells to become specialists.

Over the course of time, as flagellum-whipping and nutrient-collecting cells developed an ever-closer relationship, a new specialization probably arose between the whipping and collecting cells that really sealed the partnership. Rather than help with water movement or nutrient collection, these new specialists most likely played a role in helping to transfer nutrients more efficiently from nutrient collectors to whippers in a different location. This would have freed flagellum carriers to be in ideal locations for creating currents, rather than always being stuck directly next to the nutrient collectors. Such freedom would have allowed the early multicelled cooperative communities to build structures that improved the dynamics of water currents and led to increased nutrient collection.

Of course, the cells that transferred nutrients from collectors to whippers would also need some nutrients themselves. They probably functioned a bit like mercantile middlemen, taking a cut of the nutrients that they transferred. And, as they took this role on, evolutionary forces would have led the offspring of these transfer cells to give up carrying flagella and lose their nutrient-collecting machinery. It was no longer needed.

The teamwork created by so many specialized cells developing and functioning together would have made it possible for such colonies to outcompete most single-celled organisms in the surrounding environment. Cooperation favoured survival, and this evolutionary pressure would have been of such intensity that the connected cells would have become so specialized that they would eventually leave behind their status as individual animals and become cells entirely dependent upon one another. The flagellum carriers would become dependent upon other cells to feed them, while the hard-working nutrient collectors would have to depend entirely upon flagellum carriers to direct food towards them. And so from many single-celled organisms benefiting from cooperation came one animal.

Reproduction in these early animals was a simple matter of realizing individual cells that had not yet developed any specializations. Since all cells in these animals descended from a single mutant parent cell with offspring that stuck together, all cells carried the genetic information to develop the physical form to take on specific tasks. As a few reproductive cells settled in an area to form, a new animal started to divide and multiply. Some would take on the role of creating a current and develop flagella, while others would take on nutrient

The common jellyfish, also known as the moon jellyfish, is found in waters throughout the world, mostly near coasts. The earliest jellyfish were probably similar to sponges.

collection and develop characteristics suited to the chore.

When we describe this evolutionary process, it is easy to slip into language that suggests that these individual single-celled organisms actually knew what they were doing, and that they came together through an understanding that mutual cooperation was best. But single cells, whether the cells of the sponge or neurones in the human brain, are not intelligent.

Instead, we know from work done by researchers like Degnan that, as modern sponge cells divide and grow, they send signals to one another, and these signals control what sorts of characteristics the individual cell develops. Cells near the exterior of the sponge where water currents need to be generated will tend to develop into flagellum carriers; cells towards the centre of the sponge structure will develop into nutrient collectors. This means that body plans for animals are inherently built into their development.

With several hundred million years of evolution behind them, modern sponges have a little more complexity than the hypothetical early animal whose appearance we have just witnessed. The sponges you might pull out of the surf on holiday also have structural cells and digestive cells in addition to flagella, nutrient collectors and nutrient transferrers, but they are not so different from the fossilized sponges that date back to that early post-Snowball Earth era.

Degnan's sponge-reforming experiment shows that sponge cells are unique in how each cell can survive independently and that they hark back to those early days when cells were just starting to specialize but had not yet become totally dependent on one another for survival.

Fossils that are clearly recognizable as sponges have been found in rocks 570 million years old. There are even some finds suggesting evidence of even earlier sponge evolution, though the fossil evidence is not as clear. Either way, sponges that looked like those alive today evolved during, or shortly after, the Snowball Earth period and were thriving in ancient waters.

——

The gently pulsating body of a jellyfish may look beautiful in wildlife films, but that admiration is short-lived if you meet one washed up on the beach, its squelching translucent gloop buzzing with flies. An encounter with one in the sea is even less pleasant if you get a tentacle slashed across your arm or leg.

Unlike sponges, you will not want to find a jellyfish in your bath, but their basic biology suggests that, when they first evolved, they were not so different from the sponge. Moreover, fossils of these animals, collectively known as cnidarian, and dating to at least 560 million years ago, tell us that they evolved around the same time as sponges.

In modern oceans, cnidarians use specialized stinging cells called nematocysts that hang down below their bodies in the water. Although

Sponge releasing sperm into the water. Sponges are primitive animals whose bodies contain no muscle or nerve cells. They are supported by a mineral skeleton embedded in a gelatinous matrix.

they do not actively hunt for prey, the delicate yet impenetrable forest of spindly threads is more than a match for small fish. The stingers paralyse or kill their unsuspecting victims, which the jellyfish's body then engulfs and slowly digests.

Jellyfish are relatively simple animals. Like sponges, they don't have brains. However, some do have rudimentary nerve networks that allow them to react to their surroundings, and many jellyfish can coordinate the contraction of their cells to swim around. They can also move in the direction of smells they detect in the water and even manipulate food items that they catch with their tentacles.

Cnidarians come in a great many varieties. While jellyfish and their relatives are the best known, anemones are also part of the group. These flower-like animals creep along the sea floor and use their stinging tentacles to capture small animals in their path.

Corals are also cnidarians. If you pick up a piece of coral washed up on a beach, it looks dead and like a white, pockmarked stone. That is because it is dead; what you hold is the remains of the coral's home. Each tiny hole in a piece of coral would have held a very small, anemone-like animal that can sting and capture small prey. Although their food is always quite small, the sting of some corals, collectively known as fire corals, can be extremely painful to divers who accidentally brush up against them.

———

However, in their most basic form, cnidarians are simple capsules made of cells that capture nutrients. Unlike sponges, which have many holes in their bodies for water to be filtered through, cnidarian bodies are made of sheets of cells that are relatively impermeable.

Their evolutionary story would probably sound like the sponge's: a single-celled organism that did not entirely separate from its parent cell found that by sticking with another single-celled organism it could live and reproduce more successfully than if independent. Two joined cells create more surface area for capturing food so the sheet of organisms survives and grows as it reproduces. Over the course of time, the different organisms start to specialize, just as we saw in sponges.

Some of these single cells became specialized at disabling and catching prey, ultimately evolving into the stinging nematocysts common in all cnidarians today. Other cells became specialized at digesting prey or transferring digested nutrients around the network of connected cells in the collaborative sheet.

Ultimately, like sponges, the single cells in the collaboration lost their ability to survive on their own; the sheet was transformed from a collective mat of organisms into a single animal.

Sponges and cnidarians are the great survivors from a pivotal moment in the evolution of animals. No longer would life in the oceans be limited to single-celled organisms. Now complex life forms would start developing, heralding the beginning of animal life as we know it.

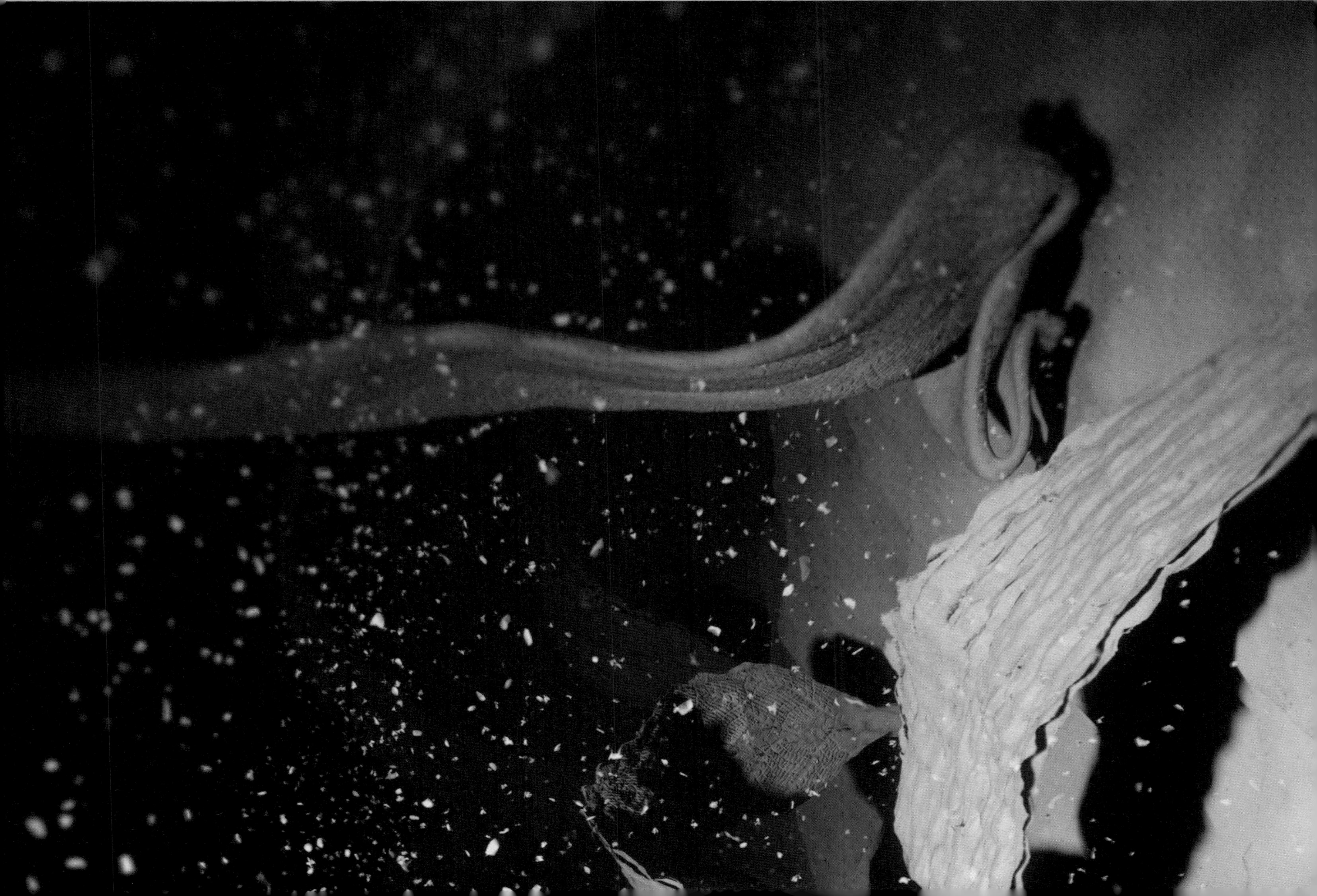

THE OLDEST ENGLISH-FOUNDED settlement in North America is believed to be the city of St John's, located on the island of Newfoundland in Canada. But a 100-mile-drive south along the coast takes us back to an ancient time when the very first animals on the planet were evolving. Here, on a rugged swatch of coastline, lies one of the richest fossil beds in the world. It is called Mistaken Point.

The site was given its curious name because ships sailing along the coast would become disoriented in the fog that blanketed the area. Drifting off course, they would crash spectacularly into the black rocks at the base of the cliffs. Today, the coast is still treacherous, but the area is now far better known as a fossil wonderland than for a graveyard of ships.

Hundreds of millions of years ago, Mistaken Point was at the bottom of the ocean, and the fossils found there are the world's only deep-water fossils from the post-Snowball Earth era. Frequent volcanic eruptions dumped tonnes of ash into the sea, which rained down onto the sea bed. The soft-bodied animals dwelling in the waters were buried alive by the ash. Unable to dig themselves out, these boneless animals rotted away and the impressions of their bodies that they left behind help to tell the story of the earliest colonization of animal life.

'The volcanoes were very important to the preservation of the fossils at Mistaken Point,' explains palaeontologist Guy Narbonne of Queen's University, Ontario. 'Every time they went off, they would fill the ocean with volcanic ash, which would smother the creatures on the sea bottom. This had a few results. The first result is that it preserved them perfectly. Secondly, the ash killed them in their own habitat, perfectly preserving their form and location. The fossils can then be studied using modern ecological techniques to deduce their behaviour and interactions while alive. The third result is that the ash provides information to accurately date the fossils. Volcanic ash contains a mineral called zircon, which contains uranium. Uranium is a radioactive element, which decays to lead within a known time period. So by studying the ratio between uranium and lead in the zircon, we can precisely know the time at which the volcano erupted and thus date the fossils on the sea bed.'

The fossil site itself is incredible. There are more than one hundred layers of rock that sit one on top of another like sheets, each representing volcanic events that took place between 575 million and 560 million years ago. In different areas of Mistaken Point, the sheets of rock have become exposed by weathering from wind and rain. Some of these exposures resemble large, slightly angled tennis courts, smooth, flat and full of fossils.

But Mistaken Point's undersea world today is a far cry from how it may have looked 575 million years ago. For a start, it would have been pitch-black down there, with parts of the ocean bed without light for many millions of years. It is, however, likely that a gentle current flowed through the waters, and this is probably what allowed life to thrive. But the animal life forms that inhabited this ancient ocean were extremely unusual.

'Even though they were in water so deep that no light could reach it, it was teeming with life,' Dr Narbonne explains. 'This would be very strange life to us. This is not like anything else that exists today. The creatures were immobile. Nothing had a mouth or muscles. There was probably an eerie, whitish colour to everything. Creatures died where they lived, and their bodies were constructed in a way that we don't see today.'

**Mistaken Point in Newfoundland is one of the most important fossil-bearing sites in the world.**

The rugged coastline of
Mistaken Point was given
its name because ships sailing
along the coast would become
disorientated in fog.

Ancient sponges and cnidarians attract a lot of attention today because they belong to animal groups that are still alive. Their modern relatives grant palaeontologists great insight into how their fossilized ancestors are likely to have behaved. But sponges and cnidarians were not the organisms that dominated the ancient oceans. They had plenty of company in the form of strange-looking organisms that we know very little about. Fossils of these long-extinct animals show us what these creatures looked like, but their biology is barely understood.

The mystery surrounding these long-extinct animals stems from the fact that they do not look like any life on the planet today. Humans have something very important in common with one another, and with dogs and cats, and even with *Tyrannosaurus rex*: we all have bilateral symmetry – that is to say that the left and right sides of our bodies are mirror images of each other.

Such symmetry was rare in the ancient waters where multicellular life arose. Cnidarians show radial symmetry, meaning their bodies look something like well-organized pizzas and can be cut into identical wedges. Some cnidarians show three-part symmetry and can be cut into three identical pizza slices, while others show four-, five- or even six-part symmetry.

Sponges have no symmetry, and are often totally amorphous. The biology of their distant ancestors, however, makes them easier for us to understand today.

If you were to swim through the seas of post-Snowball Earth, you would see some alien-looking life forms with bizarre body shapes. These are not bilateral, radial or even amorphous. Rather, they display alternating branching features that resemble fractal patterns. Microscopic analysis of their fossils reveals that each individual element of these organisms is finely branched at every scale, all the way down to

A *Charnia* fossil at Mistaken Point.

hundredths of a millimetre in diameter. Think of them like a tree with many branches, with each branch a scaled-down variation of another. These creatures dominated the ancient oceans of Mistaken Point.

'If we look at the animals we're most familiar with, mammals, birds, amphibians or even humans, we see that the left side of the body is a mirrored repetition of the right side,' Dr Narbonne explains. 'We have a plane of bilateral symmetry that passes through our face, down the middle of our torso, dividing us into two mirror-image halves. This is typical of most higher animals. As we trace animal life back to the more simplified organisms, we pass through a group called the radial animals, which includes jellyfish and sponges that do not have this symmetry. The body structure of these animals is based on a circular (or radial) structure. But the branching structure found in the post-Snowball Earth seas are yet further removed from these in that their bodies are configured as a fractal structure. This is multiple branching on multiple scales, which is neither radial nor bilateral and differs completely from the two other main groups of animal life.'

The first recorded discovery of this strange, fractal life took place in Australia in the Flinders Ranges, north of Adelaide, in 1946. Reginald Sprigg, an Australian geologist, discovered ancient fossils that had odd branching patterns, though his findings gained little attention.

Instead, it was Roger Mason's discovery of *Charnia* that enabled the next great leap in our knowledge of animal evolution. As we saw in Chapter 1, Mason, now a geologist at the China University of Geosciences, went rock climbing in 1957 in Charnwood Forest near Leicester with his friends Richard Allen and Richard Blachford and discovered a strange, frond-like fossil. This fossil, which was incontrovertibly believed to be Precambrian, was studied and named by geologist Trevor Ford. Debate quickly swirled around what *Charnia* was because it was unlike any other group of living things anyone had seen. It slightly resembled a plant – perhaps some strange kind of fern – but it also looked like a modern cnidarian animal called a sea pen.

———

Over time, Mistaken Point has emerged as one of the most important fossil-bearing sites in the world, with simply wondrous and scientifically revealing finds. To date, more than 200 *Charnia* fossils have been found there (along with countless other creatures), formed just after the animal kingdom branched off from the rest of the living world.

'The fossils of Mistaken Point show the origin of large multicellular creatures,' Dr Narbonne says. 'This is when life began to get big. It tells us a lot about the conditions that led life to increase in size. It also reveals a unique experiment, you could say a failed experiment, in Earth's evolution.'

# Mistaken Point

"It's been a long-held ambition of mine to visit Mistaken Point in Newfoundland, Canada, and see the exquisite fossils there, which I managed to do for the first time whilst filming *First Life*. A unique place, it is one of the most important fossil-bearing sites in the world. The ancient rocks along this coastline span 10 million years of Ediacaran fossil history in more than 100 layers of rock.

By examining different layers of rock here it's possible to virtually travel in time from a period when *Charnia* existed alone in the depths, through millions of years of evolution, to rocks that contain a far greater collection of curious Ediacaran characters, quite different from anything you might see on Earth today. Each one of the layers of rock was once mud lying at the bottom of an ocean – a deep, cold ocean that was almost certainly pitch black.

For most people, a lifetime or 100 years seems an incredibly long time, which makes 570 million years simply unimaginable. What's even harder to comprehend is that fossils of these creatures, which must have had the most fragile composition, have survived all this time.

Everywhere you look at Mistaken Point there are markings in the rocks of one kind or another. It's just as if children have been playing in wet sand. The sheer number of organisms preserved in some of the layers gives you the feeling of walking through a carpet of ancient creatures. It's a magnificent place, and one I feel hugely privileged to have seen."

Fossil beds from the same time period were also discovered in South Australia, Namibia and the White Sea coast of Russia. These finds showed that *Charnia* was extremely common in the ancient seas between 575 million and 543 million years ago.

But careful comparison of the *Charnia* fossils revealed that there were variations in the patterns, which tended to be localized. Many palaeontologists speculated that these were populations of different species of frond-like organisms of which *Charnia* was just one example.

To our modern eyes, conditioned to left–right bilateral symmetry, how weird and wonderful a stand of these fern-like animals would seem. You might encounter *Arborea*, with branching formations that look like peas in a pod. At just below knee-height, is the majestic *Swartpuntia*, with branching tubes on its fronds that spread out from a central stem like a whorl of miniature-veined sails. These shapes and structures differed strikingly from the zigzag patterns found on *Charnia*. But why should they differ at all?

Palaeontologists are born to compare. It does not matter if the fossil is 5,000 or 5 million years old; when palaeontologists are not sure what they are looking at, they hunt for comparisons in museums. There, they search among all the plant, animal and fossil specimens for something similar. They hope the answer is waiting, long forgotten, in the bottom of a dusty box in an archive storeroom.

If they find a match, they can identify the new fossil as a member of an existing species. Without a match, they carefully consider which specimens look vaguely similar and speculate where on the tree of life the new fossil might belong. This is straightforward if the fossil is suspected of being an ancient rodent, for instance. If it has the long, continuously growing incisor teeth common to all rodents, then it is safe to put the new fossil as a member of the rodent family. And it is hardly a leap of faith to look at the behaviour of rodents living today and conclude that the fossilized species probably behaved in a similar fashion.

However, if fossil animals have no living relatives remotely similar, it is much harder to work out how they lived. Take the horned dinosaur *Triceratops* as an example. The only descendants of the dinosaur lineage alive today are birds, but none of them has sharp, metre- (3-ft-) long horns growing from their heads. There are not even any living lizards or snakes, which are distant cousins of the dinosaur lineage, that have horns structured like those of *Triceratops*.

So what were those horns for? Perhaps we can glean clues from other horned animals around today, such as elk, deer, goats and buffalo. Even though these animals are mammals – and, therefore, not even closely related to dinosaurs – they can at least give palaeontologists an idea of how animals with horns like *Triceratops* might have behaved.

Follow a herd of buffalo as they graze the African savannah.

Inevitably, a hungry pride of lions comes prowling by, looking for their vulnerable young. But the buffalo have strong protective instincts, and they gather their young into the centre of the herd and face out towards their predators. The lions are now confronted by a solid wall of bovine anger and lowered horns.

Observing this kind of behaviour, palaeontologists suspect that *Triceratops* behaved similarly, bunching together to form a wall of lethal spears when danger came near.

It is also suspected that *Triceratops* males used their horns to fight for breeding access to females, just like deer rut in the autumn. Recent research backs this up with findings of scratches on *Triceratops* skulls that look like they were made by horns. The most feasible explanation is that two dinosaurs rammed one another at high speed.

But *Charnia* and the other bizarre frond-like organisms from the post-Snowball Earth age present a much greater challenge for palaeontologists. If ever fossil collectors needed Sherlock Holmes's powers of deduction, it was in trying to solve the mystery of *Charnia*.

Let us start with the most obvious suggestion: that fossil fractal frond specimens look like the leaves of plants, so perhaps they were actually plants. Given the physical similarities, the idea is appealing, but it is almost certainly wrong. We know from studying ocean-dwelling plants like kelp that the best way for them to capture sunlight in order to photosynthesize is to have many leaves and very long stems. The longest *Charnia* specimens are just a metre (3 ft) in length, and most are much smaller. They also all seem to have just one leafy structure at the end of their stems. These two characteristics alone hint that these organisms were not gathering sunlight.

**Examining a *Charnia* fossil exposed at Mistaken Point. It was preserved by the volcanic ash that smothered the creatures on the sea floor.**

We know that in the depths of modern oceans light does not penetrate and plants do not grow. We also know that while life changes over time, the laws of physics do not. If light does not penetrate deep water today, it would not have done so 560 million years ago, which eliminates the possibility of *Charnia* using photosynthesis. So *Charnia* may look like a plant, but it is undeniably an animal.

Nevertheless, plants may still give us clues about how *Charnia* behaved and why it developed a fractal pattern. The leaves of a plant do more than collect sunlight. They actually 'breathe' in gases from the air around them – not with lungs but with small cells on the leaf undersides that can open and close. Leaves allow gases to travel easily in and out of the plant because they are thin and flat, with a large surface exposure to the surrounding environment. If, instead, plant leaves were thick and rounded with numerous cells hidden from the surface by many layers of surrounding cells, there would be much wasted tissue on the inside of the plant that would be cut off from the gases in the surrounding environment.

So it is quite possible that the fractal frond structures were being used to absorb minerals or gases dissolved in ocean water. Researchers who have explored this possibility have searched for chemical signatures in fossilized sediment to see if they once grew in water rich in chemicals. (Remember that living organisms are likely to have started by first consuming chemicals that were released deep within the Earth, and to this day there are still bacteria living by deep-sea vents that do this.)

Formulating a theory is easy; finding the evidence in this case is more difficult. No fossil vents or chemical seeps appear near sites where these fractal frond organisms were dwelling, nor any chemical signatures in the surrounding sediments.

So we have to look at other organisms with frond structures. There are plenty to choose from, including fungi and some animals, although the purposes of the frond are wildly different in each case.

Mushrooms are called fruiting bodies by scientists, and the structures that fungi produce are entirely designed for reproduction. They release spores into the air that drift off to other members of the same species and allow the fungi to sexually reproduce.

Could the *Charnia* fronds have distributed reproductive material? It is certainly possible, although some palaeontologists argue that the fronds would have been rounded and bulbous like mushrooms if they were for reproduction.

Where else can we look? Let us return to an animal we mentioned previously, the sea pen. Species of sea pen are found in warm, tropical waters and in cooler, temperate seas. Their colours range from pastel pink and an enigmatic violet to vivid yellow. In many ways, they look like delicate, feathery quills sticking out of the sea floor.

To the casual observer, sea pens are not animals at all. They do not

# Mistaken Point's Fossils

" It always seemed to me a bit curious that Precambrian fossils are as scarce as hens' teeth throughout most of the world, yet are so very abundant along the Avalon Peninsula coastline of Newfoundland. What makes this isolated area so special?

The answer: volcanoes. When these fossils were created this was a much more hostile place. Massive eruptions rained millions of tonnes of ash onto the sea. The ash sank to the bottom, enclosing everything living beneath it like a sub-marine Pompeii. Over millions of years the ash itself was buried under sediment and squeezed under the pressure, turning it into rock. Over an even greater length of time, hundreds of millions of years, colossal tectonic forces thrust the whole sea floor upwards to its current position on the coast of Canada. The level of radioactivity of a volcanic element called zircon in the rocks allows scientists to date the eruption, which created the ash layers surrounding the fossils rocks, to precisely 565 million years ago.

A lot of the Mistaken Point fossils have funny nicknames, such as 'feather duster' and 'spindle'. Their forms are so different to modern animals that the palaeontologists who first found them decided to name the fossils after objects that they resembled. My personal favourite is the 'pizza disc', *Ivesheadia*. It's called a pizza disc because of its circular shape and surface pustules, which resemble melted cheese. The pizza disc was a very simple creature, living flat on the ocean floor, and the only one of its kind. It was also one of the first comparatively large animals to inhabit our planet, all those millions of years ago."

*Ivesheadia*: the so-called 'pizza disc' is one of the creatures that once lived on the ocean floor now preserved at Mistaken Point.

appear to have any moving parts, and they are rooted to the rock by a solid stem. However, sea pens are cnidarians, and thus related to jellyfish and anemones. They grow in areas where natural currents are present and use their stinging cells to capture tiny organisms drifting by.

Could the fractal frond organisms have been ancient cnidarians that behaved like or even evolved into sea pens?

Shortly after Roger Mason discovered *Charnia*, this is precisely what palaeontologist Martin Glaessner of the University of Adelaide argued and, to this day, many palaeontologists agree. He reasoned that even if the frond organisms did not yet have the specialized stinging cells, they could have simply been capturing organisms in the water.

However, as with other theories, there is conflicting evidence. Modern sea pens have tentacles that can be seen if the animals get buried in fine sediment. Indeed, fossils of recently living sea pens often have clear tentacle marks. But even though the frond fossils of Mistaken Point were preserved in the finest of sediments, ash, there is no evidence of tiny tentacles.

Of course, just because tentacles are absent from the fractal fronds in the fossil record does not mean that tentacles were absent on the fronds when they were alive. They could simply have been too small to be identified, or were destroyed before preservation. Scientists have experimented with cnidarians to see how quickly they rot in different settings. The results of their tests suggest that if these frond organisms were cnidarians and had tiny tentacles, these tentacles would have been preserved as fossils only if the animals had been buried entirely in sediment within a couple of days. Had the sediment rate been slower, scavenging organisms, like bacteria, would have got at the frond's tentacles and rotted them away.

With every theory exposed to heated debate, the argument that tentacles might be absent from fossils because of decay is also disputed. Again, the evidence comes from sites like Mistaken Point, where the fractal fronds have been encased in volcanic ash or storm-washed debris. We know that these forms of sediment can be deposited rapidly. Tonnes of mud can be dumped in a flash flood and, in some areas where volcanic eruptions are taking place, ash layers many metres thick can come down in less than 24 hours.

For *Charnia* or other fractal frond fossils that were created from ash or storm sediments in depositional environments, the argument that tentacles were rotted away is somewhat weakened.

In the absence of conclusive support from modern organisms with fronds, palaeontologists have proposed an alternative theory. Perhaps the ancient frond organisms were just absorbing nutrients that were already present in the ocean water. In this case, the fronds would not need tentacles, and they would not need to live near chemical seeps or underwater volcanic vents if nutrient loads in the ocean as a whole

Reconstruction of *Charnia* alive on the sea floor.

were already high enough to sustain chemical processes that kept alive fractal fronds like *Charnia*.

This theory suggests that the fronds engaged in a chemical exchange similar to the way in which plant leaves exchange oxygen and carbon dioxide. This helped them to swap nutrients efficiently, along with the oxygen that they were probably drawing from the water as it flowed by. The idea of the fronds collecting nutrients from flowing water also explains their structure. With their rigid stalks and single, somewhat teardrop-shaped fronds, the organisms would have been well suited to standing upright in currents, holding their position and collecting nutrients.

But why were there so many different frond types? Were the various patterns helping the fronds to collect food in different ways and therefore allowing them to survive in different environments?

—

*Similarities in the natural world can be deceiving. Under the right conditions, evolutionary forces can drive organisms that are not closely related to look similar.*

—

Dr Narbonne believes that the variety exists because the frond organisms all evolved their mysterious way of life independently, a process called convergent evolution. As a simple rule, if two animals look similar, they are likely to be related. Rats and mice, ravens and crows, and even chimpanzees and humans are all good examples of this principle at work. These animals look similar because, in the recent past, they were one species, but through new traits appearing in the population, the single species split. However, this rule does not always apply; similarities in the natural world can be deceiving. Under the right conditions, evolutionary forces can drive organisms that are not closely related to look similar.

Consider a hypothetical population of land-dwelling predatory mammals living near a river filled with fish. If the mammals are mainly hunting small terrestrial animals like rabbits, they will most likely be good runners, have flesh-cutting teeth, and keen senses for detecting their prey. Perhaps they look a bit like a cat. For some reason, this creature breeds prolifically and soon the population has become too large. With lots of predators and not enough prey, starvation sets in. Some predators die before being able to breed. Under such circumstances, some of these desperate feline predators might make a stab at eating the fish from the river's shallow waters. Any success at this would reward them with nutrients that might help them survive long enough to breed.

Over time, characteristics arise from sexual diversity or mutation that enhance this fish-feeding ability so that the populations of predatory mammals do not have to compete with their own kin for the dwindling supply of rabbits available in the forest. For the sake of this explanation, it is irrelevant what the advantageous traits might be – all that matters is that the traits give the predators an edge when it comes to hunting fish.

Eventually, if fish feeding were to become a regular activity and the predators were to spend more time in the water pursuing the fish, those with traits best equipped for movement through water, such as streamlined bodies, a complete lack of fur, and fins instead of long limbs, would eat the most, survive the longest and breed the best.

After hundreds of thousands of years of this evolutionary process, those forest-hunting land mammals that started out looking like cats would begin to look a lot like fish – complete with fins, smooth, hairless skin and streamlined bodies. Such a resemblance would have nothing to do with any evolutionary relationship to fish, but would instead be a direct result of moving into the watery environment that they now share with fish.

In other words, environments shape animals in a way that complies with the laws of physics. The ability to move more efficiently through water and a streamlined shape are incredibly important because they cut down dramatically on resistance. For an animal to achieve any measure of speed underwater, all characteristics that appear in the population that improve its smoothness or its streamlining will be selected by evolutionary forces. This is why penguins, seals, dolphins and fish all have similar bodies. They can be said to have 'converged' on the same form.

Environments are not the only things that cause animals to be similar and to solve a problem independently. The animal's behaviours can drive this effect as well. Large eyes tend to be common in animals that are active at night, robust teeth are often found in scavenging animals that break bones and suck out nutrient-rich juices, and long legs are frequently seen in sprinting creatures.

So with convergent evolution in mind, Dr Narbonne theorizes that the fractal frond organisms may not be related at all. Their fossils may look alike but only because they all behaved in a similar way. If nutrient-rich waters were flowing in strong currents around 580 million years ago, it is possible that, in the same way flippers and streamlining are important for swimming organisms, stalks and frond shapes were the ideal structures for whatever process of feeding these fractal frond organisms were engaging in. He and his colleagues argue that completely unrelated lineages of early organisms experienced convergent evolution to become similar in form and take advantage of the ocean's riches.

A key date for the rise of animals was 580 million years ago. It is

Excavation work in China has
uncovered fossil embryos – tiny
sand-grain-like specimens.

when the first multicellular life noticeably appears in the fossil record, but Dr Philip Donoghue at the Swiss Light Source in Switzerland is conducting work that suggests animals arose a lot earlier. He and his team argue that they have evidence showing animal life appearing around 630 million years ago.

Dr Donoghue studies grains of sand that are smaller than a millimetre in diameter. They are from a vein of rock in southern China that appears to have the same composition as bone. For decades, sediment from this vein of rock has been commercially valuable as fertilizer. But Donoghue finds it valuable because the grains of sand are actually ancient tiny fossils.

Animals are formed from the union of genetic material from two separate parents, most commonly in the form of a sperm cell and an egg cell. This merger occurs and creates a single cell, which then begins to divide (replicate). First, it divides into two, and then each of the two new cells divides as well, leaving four cells. Next, these divide to create eight cells, and then sixteen, thirty-two, sixty-four and so on.

We know what these dividing balls of cells, called embryos, look like because we can bring the sperm and eggs of animals, including humans, together in the laboratory and look at the resulting cell divisions under the microscope. Using a series of microscopes, Dr Donoghue has scanned the surfaces and probed the insides of these Chinese sand-grain-like fossils. He concludes that you can see structures that look remarkably similar to the dividing balls of cells found in modern animal embryos.

The fossil embryos are simple, however, and do not seem to have any specialized cells or clumps of cells that form anything like the unique tissues seen in modern animals. But even just balls of cells are incredibly exciting because they prove that multicellular organisms were growing in the very earliest post-Snowball era seas. Donoghue is not certain what animals the balls of cells would have eventually grown into, because, during the first hours of all animal development, the balls of cells all look the same, regardless of whether they will develop into humans or sponges.

Why the embryos came to be fossilized in a single layer is also a mystery. Dr Donoghue speculates that, like some animal embryos today, the embryos of ancient organisms would sometimes go into a state of arrested development, called stasis, as they waited for conditions like temperature and nutrient availability to reach certain levels. If large numbers of embryos were released in some sort of grand spawning event similar to that of corals today but conditions took a turn for the worse and forced them into stasis, they might have collected and fossilized in a single location and thus become the fossil vein seen today.

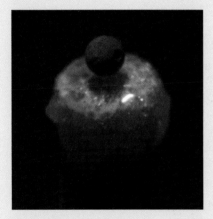

Fossil embryo of *Markuelia*, thought to be the first animal to evolve a complete digestive system. This specimen was scanned at the Swiss Light Source.

In our exploration of the birth of complex life on Earth, we have looked at the fossils of many different individual animals. In a sand-like vein in China, more than 600 million years old, we have glimpsed dividing embryos of perhaps the very first multicellular animals on this planet. We have also discovered fossils that bear remarkable resemblance to sponges, jellyfish and corals, and we have encountered strange fractal organisms, such as *Charnia*, whose behaviour and life cycle we can only imagine. But studying early life involves more than peering at fossils. Palaeontologists also want to discover how different species interacted with one another, and to enable them to do this they must employ the science of ecology to understand how these fossilized animals behaved in their surroundings.

To understand more about these interactions, we must return once again to Mistaken Point. Here, 550 million years ago, a great diversity of primitive life was thriving. From the tall frond-like organisms similar to *Charnia*, through smaller bush-shaped organisms, to dainty spindle-shaped organisms on the sea floor, the variety of shapes, sizes and designs found in the fossil record there is astonishing. While debates ensue about exactly what these different organisms were feeding on, there is near universal agreement that they were depending upon food being brought to them by the water that was slowly flowing by.

'What would life be like for these earliest animals at Mistaken Point?' Dr Narbonne contemplates. 'For the greatest part, it was a pretty good place to live. There was no sun, but it was a quiet environment for the organisms that lived in the deep water, with the gentle currents bringing them all the food and nutrients that they could possibly want. However, occasional earthquakes or great storms would unleash masses of sediment, and this would flow down in a turbid mixture to annihilate the community – forcing it to move – or, worse, wipe it out altogether.'

According to Dr Narbonne, one of the most remarkable characteristics of the Mistaken Point ecology is how different organisms found their niche at different depths. This division of an environment into sectors is seen in ecosystems today. In forests, for example, tall trees absorb direct, intense sunlight, smaller trees survive on less intense sunlight that filters through the tops of tall trees, and shrubs on the ground tolerate shade. Palaeoecologists suspect that the Mistaken Point organisms were doing much the same thing. With nutrients being carried by ocean currents instead of sunlight, there were tall fronds, medium-sized shrubs and ground-covering organisms, each collecting nutrients from different parts of the water column.

The Mistaken Point fossil treasure-trove may also help the palaeoecologists answer the burning question of reproduction. Fossils rarely catch animals in the act of breeding, and how the fractal frond creatures living at Mistaken Point reproduced remains an unanswered question. There are two intriguing ideas currently under consideration.

The first makes use of the fractal feature of the fronds. The Mistaken Point fossils show that the branches and elements on large fronds are exactly the same as those branches on smaller versions of the organism. This suggests that perhaps the frond-like organisms detached bits of themselves; these would then take hold, divide and grow as unique individuals, following the fractal blueprint. This is seen today in many plant species and is widely known as vegetative reproduction. However, if this were the only method of reproduction that they used, we would expect to find large groups of nearly identical frond-like organisms within dropping distance of one another. But a little time spent lying on the ledges of Mistaken Point shows that this is not the case.

Dr Narbonne prefers another explanation. He argues that fractal frond organisms have been found all over the world, and in order for them to have spread so widely, vegetative reproduction could not have been the only method used to reproduce. It is simply too slow and does not distribute offspring far enough.

Instead, it is possible that these fractal fronds did something similar to what corals do today. They may have released large quantities of sexual cells called gametes (sperm and eggs) into the water. Corals of the same species are all tuned to the length of day and temperature, and these cues mean that all the corals in an area release their gametes at the same time when the right conditions are met. It is an amazing sight to see – the waters suddenly cloud over as millions, perhaps billions, of coral gametes are set free. Since each coral organism releases its gametes simultaneously, there is a good chance, even in the vast ocean, that some sperm will meet some eggs and fertilization will take place. The worldwide distribution of the frond-like organisms hints that they probably did something similar. However, whether they used only this strategy or a mix of gamete release and vegetative reproduction is not known.

These fossil trails at Mistaken Point are thought to be the oldest known traces of animal movement.

As research goes on, so Mistaken Point continues to release its secrets. Until recently, the ancient organisms of Mistaken Point differed from our modern world in one important feature: their lack of locomotion. *Charnia* and its fractal fellows were firmly anchored to the sea bed. But patterns in the rocks hint at some of the first tentative evolutionary steps in animal movement. A key difference between the world represented by Mistaken Point fossils and most ecosystems today would be the occurrence of animals on the move.

The oldest-known traces of animal movement were thought to be around 560 million years old, and none was thought to be present at Mistaken Point, yet extraordinary new fossil finds made by Dr Martin Brasier, a professor of palaeobiology at Oxford University, and his graduate student Alexander Liu may be pushing life's first slithers back earlier in history.

In recent years, while exploring the rocks of Mistaken Point, Brasier and Liu have spotted over 70 different markings. Many of these marks were just a couple of centimetres in length, but some were as long as 17 cm (7 in) and 13 mm (0.5 in) wide. These markings are remarkable because they have not been gouged out of the rock in recent history through the action of wind, rain or erosion by sand. Nor are they simply cracks. On the sides of every mark are little ridges that stick up, as if a child with a stick had drawn a line in some sand, piling the grains up the side. Some living thing, it would seem, has dragged itself through the sediment and pushed it aside.

Of course, at 565 million years of age and in such a deep-water environment, the possibility that the early movements were created when pebbles were dragged through the sediment by currents cannot be discounted. In reality, there is no way to prove without doubt that the traces are not geological in their origin. Even so, Brasier and Liu argue that there are two characteristics that make the markings unlikely to be the result of stones caught in currents or other non-biological mechanisms.

First, the markings change direction frequently. Since most currents flow in a single direction, the markings could not have been created by stones propelled in the water current. For currents to swirl or abruptly change direction, there needs to be some unusual geology in the area. This is possible for Mistaken Point, but unlikely. Second, the markings are all extraordinarily uniform in their width. Again, if stones and currents had been responsible for the markings, the stones would probably not be perfectly round. They would roll through the sediment, creating uneven widths in the markings.

Neither of the Oxford researchers can say for sure what animals these were. They suggest that we are looking at the trails of early cnidarians, something like modern sea anemones, which inched their way along the ocean bottom. Or perhaps they were some of the first worms to creep across the planet. Again, palaeontology is at its best when new theories are derived and debated.

———

While fractal frond organisms dominate the Mistaken Point fossil beds, they are not the only strange creatures to be found in the post-Snowball Earth era. Our visit to these amazing beds also introduces us to fossils of disc-shaped organisms, which are fascinating because they display left–right bilateral bodies rather than fractal bodies. These discs have been located in numerous other rocks of this ancient age all around the world. Among the most widely found is a roundish specimen known as *Dickinsonia*, which commonly appears in Australian fossil beds that are slightly younger than those found at Mistaken Point.

Disc organisms like *Dickinsonia* all seem to have had soft, segmented bodies that became covered by sediment and were recorded as impression fossils. And, like the fractal frond organisms, these early animals also boasted a great variety of designs.

Some are oval, others are rounded-discs, and others still are ribbon-shaped. The segmentation patterns on their bodies also vary, with some showing nearly bilateral symmetry and others showing more radial tendencies. They range in size, from around a centimetre (0.5 in) to just over a metre (3 ft) in length. The largest fossil of this group of organisms is, amusingly, named *Dickinsonia rex*.

———

Researchers have made educated guesses about the origins and behaviours of these organisms. Numerous studies of the fossils have suggested that they were either cnidarians or worms. A couple of scientists have even speculated that they were distant ancestors of the group of animals that eventually developed spinal cords and evolved into animals such as fish, frogs, lizards and, eventually, humans. One study has suggested that they were lichens – a curious cooperative mix of plant and fungi. Remarkably, another study claims that they have no relationship to any other living thing and probably belonged to an entirely different kingdom of life that became extinct long ago.

**With Jim Gehling in the Ediacara Hills, Australia, where many important fossils for understanding the evolution of early life have been found.**

Like *Charnia*, they raise more questions than can be answered. How did these organisms behave and live? In 1992, researchers speculated that branching structures on them were digestive organs connected to a mouth. But palaeontologist Jim Gehling of the South Australian Museum in Adelaide, who has spent years studying the fossils found in the ancient Australian fossil beds (the same layers of rock that first caught Reginald Sprigg's interest in 1946), refutes this idea.

At a symposium at Yale University in 2005, Dr Gehling and his colleagues pointed out two problems with the idea that *Dickinsonia* had some kind of prototype digestive system. First, if organs were present, why do only a few of the fossils show the branching structures associated with the proposed digestive system? Since thousands of these fossils have been found, it would be likely that many more fossils would show the impressions of these branching structures, yet they do not. Second, it would also make sense for the organ structures to be bigger and more visible in larger specimens – big animals need more food and bigger mouths – but they have never been seen in particularly big specimens. Indeed, it seems logical that something as large as *Dickinsonia rex* would leave substantial imprints of its digestive organs, but there is no evidence in the fossil that such body parts ever existed.

However, there are a couple of characteristics of *Dickinsonia* that Gehling and his colleagues argued might be clues to how the animals could have survived.

Throughout history, many different fossils have been found that overlap with one another, indicating either that the two individuals were interacting when they died, or that they died and were then somehow pushed on top of one another. With most fossils from the post-Snowball Earth period, overlapping organisms seem to be relatively common, but this is not so with *Dickinsonia* fossils. In many areas where these fossils are found, up to ten of the disc-shaped organisms are collected per square metre. Their populations in ancient days must have been in the billions, yet no two specimens have ever been found to be overlapping or touching in any way.

Gehling and his team searched museum collections specifically for *Dickinsonia* specimens that might be interacting or overlapping. They found about a dozen fossils where two of the organisms should have been touching, but one or both had deformed their normally roundish shape to avoid contact with the other nearby organism. This, the palaeontologists argue, is a clue to how the organisms might have been feeding.

Another clue was picked up by Gehling and his colleagues from strange imprints that are often found near their fossils. Close to where they ultimately fossilize, there are sometimes 'ghost imprints' that look like faint copies of the fossil itself.

Some palaeontologists have theorized that these faint copies were discarded surface tissue that the disc-shaped organisms were leaving

# Dickinsonia

"I'm lucky enough to actually have a specimen of *Dickinsonia*, which is one of the more perplexing of the Ediacaran animals. For a while some scientists thought it might have been the first organism to exhibit bilateral symmetry, but actually *Dickinsonia*'s symmetrical line is somewhat offset. Perhaps it's more of an intermediate stage leading up to the evolution of bilateral symmetry.

I found my fossil when I was filming in Australia many years ago. I looked down and there on the rock beneath me, for the whole world to see, was half a *Dickinsonia*. I couldn't believe my eyes.

As it was an imperfect fossil, I thought the geologist working with us at the site might allow me to keep it, and I was right. He let me take it home, and there it sits on a bookshelf. Every now and then I look at it and marvel, '543 million years old'. I find it a rather romantic notion that it's possible to have an object from such a remote time period sitting on your bookshelf in a room in London."

**Dickinsonia: a disc-shaped organism displaying a left-right bilateral body that provided a blueprint for more complex forms of life.**

behind. They argue that the imprints were probably a bit like the moults left by growing insects, or skins left behind by snakes. Some scientists say that the imprints are just like footprints, indentations of where the animals had once been present.

But Gehling and his colleagues have concluded otherwise. They propose that the ghost imprints are evidence of past feeding activities. Their theory holds that the disc-shaped organisms fed by absorbing nutrients from bacterial mats below them. Rather than feed with a mouth and gut system, as was proposed in the early 1990s, these organisms used their entire bodies to suck up nutrients from the bacteria that they were sitting upon. When all the bacteria mats had been sucked dry of nutrients, these vampire discs would move on to a new area of the bacterial mat where nutrients were plentiful. By feeding in this way, the organisms would have most likely left 'shadows' of their body forms on the bacterial mats, and it would have taken some time before bacteria could fill in the area where the disc-shaped organisms had been. If these 'shadows' were covered by sediment before the bacteria filled them in, ghost imprints like those in the fossil record would be preserved.

—

*Mistaken Point is an inexhaustible research site for evidence of life at the time when evolutionary change began, and one to which palaeontologists will continue to return.*

The mechanism is a lot like when you leave a doormat in the middle of a grassy lawn for a week and the grass underneath turns yellow, and then you drag the mat to a new location. The actual dragging of the doormat would not be recorded by the grass, since the mat would not sit in the travelled space long enough to kill the grass. But in the locations where the doormat sits for days, the flattened yellowed grass would take some time to recover.

What makes the idea of this foraging, nutrient-sucking *Dickinsonia* all the more interesting is that Gehling and his team's idea for the feeding tactic is known today.

Many parasitic worms that can get inside animal bodies through food, tissue infection and contaminated water have flat bodies that absorb nutrients through their skin. They do not need mouths because, inside the bodies of their host animals, nutrients are in such rich supply that they pass into the parasitic worm's body easily. Intriguingly, these parasitic worms have no need for speed nor senses, as there are no predators to worry about. So the lack of eyes and sluggish behaviour of the disc-shaped organisms suggest that they, too, lived in a predator-free world where nutrients were plentiful.

*Dickinsonia* has been found at fossil sites around the globe, but it is conspicuously absent at Mistaken Point. Gehling and his colleagues argue that this is probably due to environmental differences.

Even though the Australian fossils of disc-shaped organisms like *Dickinsonia* differ in age from the Mistaken Point fossils (Mistaken Point has fossils between 575 and 560 million years in age, while the fossils in the Flinders Ranges in Australia are aged between 560 and 550 million years), it is quite possible that *Dickinsonia* and its kin were around earlier during the period represented by the Mistaken Point fossil beds.

But Mistaken Point shows fossils from the deepest reaches of the ocean. They were living at depths of between 0.5 and 1.25 miles. If Gehling and his colleagues' interpretation of *Dickinsonia's* feeding activities is correct, then this deep and lightless environment would have been far too deep for disc-shaped organisms like *Dickinsonia* to survive. *Dickinsonia* fed from photosynthesizing bacterial mats that required plenty of light to grow. Down in the depths of Mistaken Point, these bacteria would not have been able to make their food and, therefore, *Dickinsonia* would have gone hungry. Put simply, *Dickinsonia* was an organism of shallower seas, not the dark depths of Mistaken Point.

So *Dickinsonia* and other disc-shaped organisms, as well as the fractal frond organisms, could have arisen simultaneously after the great thaw of Snowball Earth. But the fossil record is fickle. It reveals only snapshots of ancient life in different environments, leaving palaeontologists with a small fraction of the total picture. No one has yet found shallow-water fossil assemblages that date from the earlier era of Mistaken Point. But the Earth is a big place; such fossil deposits may one day be found. Meanwhile, the palaeontologists will just have to keep digging.

Mistaken Point is an inexhaustible research site for evidence of life at the time when evolutionary change began, and one to which palaeontologists will continue to return. The previous finds there have been critical because they provided the first visual testimony of what animals were like, as well as evidence of one of the earliest recorded evolutionary experiments. There, the multicellular life that lived between 575 million and 560 million years ago began to exhibit a whole new body plan called bilateral symmetry, which is the basis of you and me. Although they became extinct because they couldn't perform the more sophisticated functions of mammals, their mathematical body plan turned out to be an easy configuration for evolution to stumble across, and it eventually became the blueprint for all higher forms of animal life.

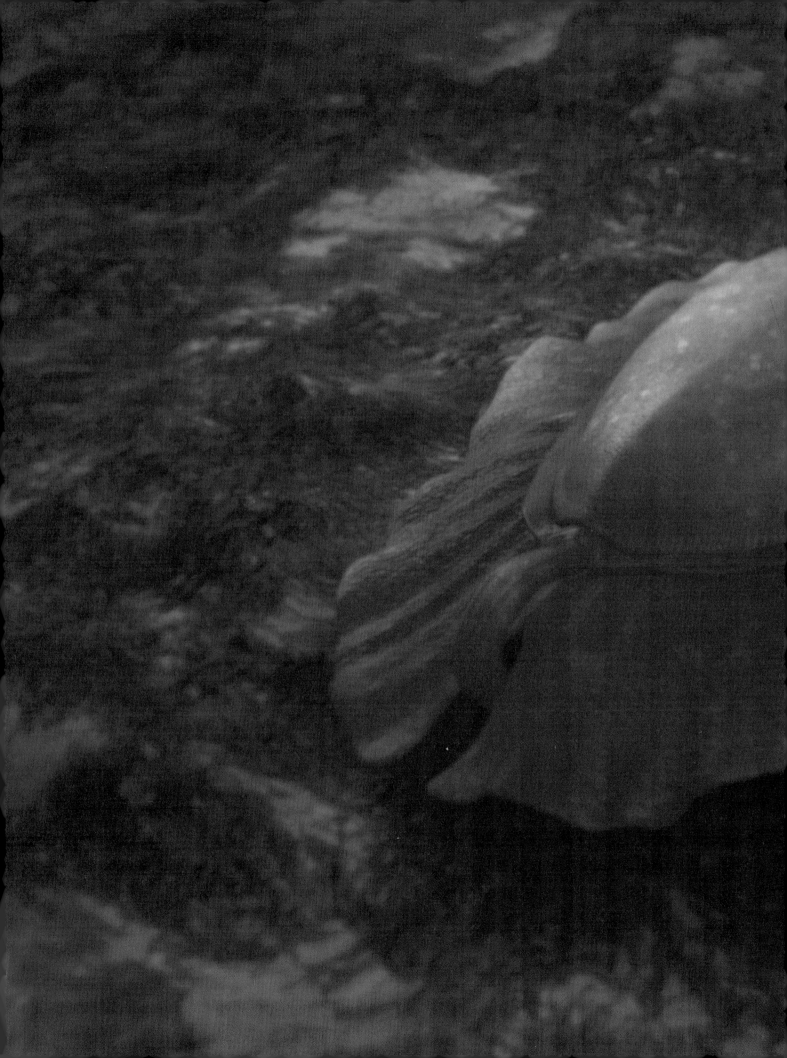

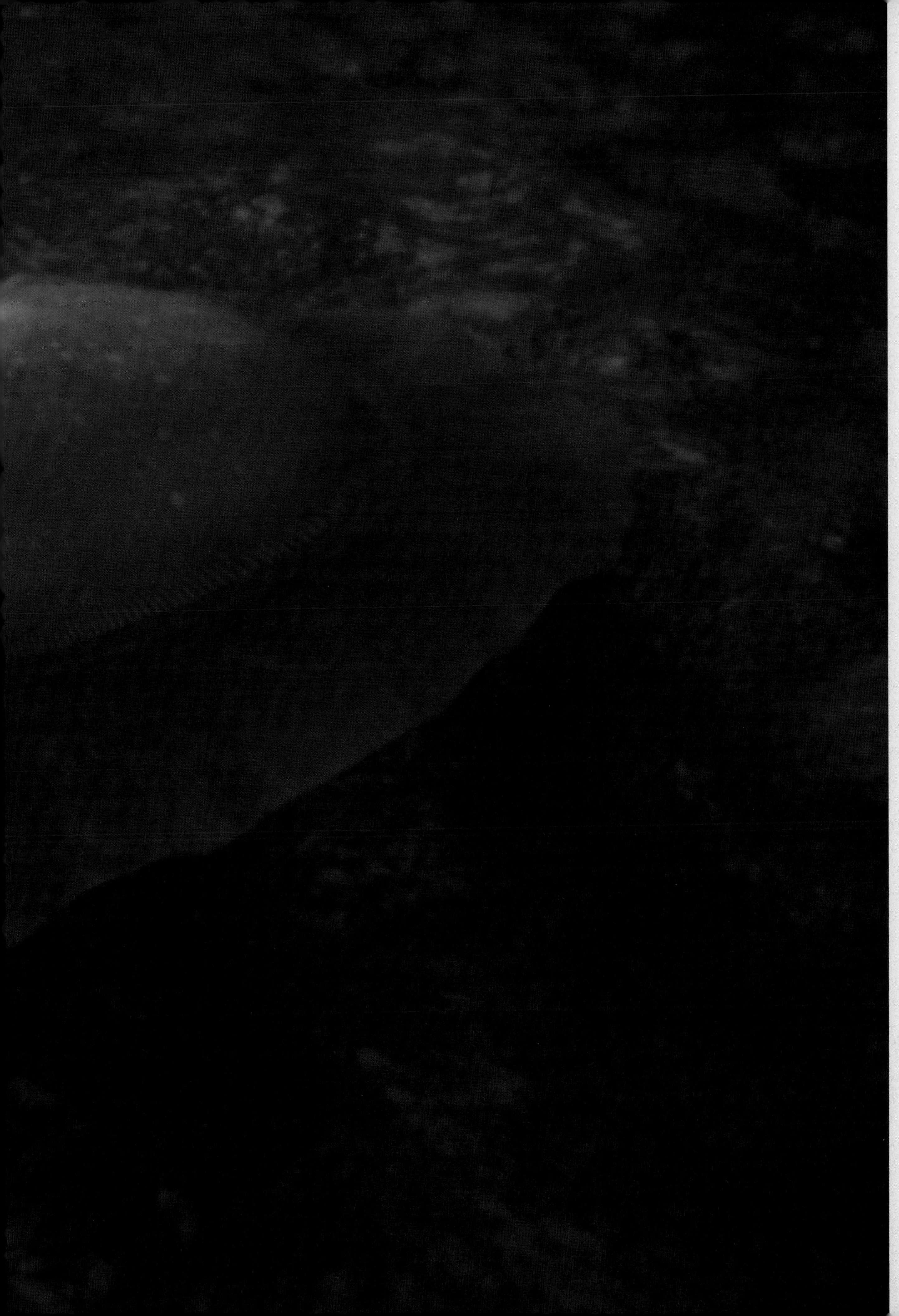

CHAPTER 6 · TO BUILD A WORM

IN THE HEART of the Australian outback, 400 miles north of Adelaide, lies a place that has played a key role in the story of early animals. It is the Ediacara Hills, in the northern part of the Flinders Ranges. In the late 19th century, this area was best known for mining, and old copper and silver mines are scattered across the landscape. But today, the most important excavations are fossils. Indeed, the geological period that these fossils came from is known as the Ediacaran, dating from 630 million to 542 million years ago.

It was in 1946 in these fossil-rich hills that Reginald Sprigg made the first discovery of bizarre fractal life that thrived in the post-Snowball Earth era. *Dickinsonia* was unearthed here, giving the area an impressive palaeontological heritage. And it is here today that palaeontologist Dr Jim Gehling and his team are uncovering the remains of a 560-million-year-old marine community that lived just after fractal animals began to die out. At this time, animals began to evolve distinct body parts and, crucially, they began to move about.

The preserved seabed in the Ediacara Hills was once a shallow reef, a different environment entirely from the deep sea of Mistaken Point. When the rocks were being laid in the form of sediment in the sea, the area would have been an amazing garden of slime. None of the animals from that era had hard body parts. The long fronds of *Charnia* and the disc-like *Dickinsonia*, soaking nutrients from a thick algal mat, would have inhabited a squelchy world on the ocean floor.

But slime does not mean simple. Some of the fossils found here suggest that *Dickinsonia* may have lived near more complex organisms – complex enough to have heads. 'The surface of the ocean floor was covered with organic ooze,' Gehling explains. 'But sitting in among that garden of slime we would have found these strange creatures.'

When Gehling takes a break from fossil-hunting and has his lunch, it is perfectly natural for him to replenish his energy with a sandwich. Evolution has conveniently located his mouth, eyes and nose all in the same vicinity. This allows him to look at and smell his food before he starts to eat it, which means that he can avoid eating anything mouldy or likely to make him ill.

It is equally useful for us to have a brain near our eyes, ears and nose. Even though the electrical signals sent by our brains travel exceedingly quickly, there is still a short lag between your nose detecting an odour and your brain deciding whether it is appetizing. By being close to the key sensory organs in your body, the brain can send signals and receive responses almost instantaneously.

Similarly, if your eyes, ears and nose were on your belly instead of your head, your ability to hear, see and smell the world would be impaired. Having these sensory organs at the top of your body allows much greater awareness.

If animals like lizards, snakes and fish – whose bodies are not upright like ours – had sensory organs anywhere other than at the front of their bodies it would be a major handicap. A snake with such an awkward arrangement would slither into walls. Equally, how would a bear coordinate its movements to track prey?

For *Charnia* and other frond-like organisms, there was no need for a head, mouth, nose or eyes. If the theory is correct, then food came to them; they did little more than absorb what was available in the ocean. The same is true of sponges and most cnidarians, so it is hardly surprising that they all lack heads.

*Dickinsonia* – that very thin creature on the sea floor which ranged in size from a penny to a bath mat – is an interesting departure from these headless organisms. Its imprints are the first evidence of some kind of mobility. We know that it had the left–right mirror symmetry of most animals with heads today but we have no evidence that it had eyes, a mouth or a nose. *Dickinsonia* most likely had some sensing cells below its body that allowed it to determine whether it was resting on a lush algal mat or not, but this is a long way from the eyes and mouths we benefit from.

But Gehling wants to find more examples of the oldest organism with a head. It is called *Spriggina* and looks like a few fish bones lined up in a row. *Spriggina* shows that in the region where *Dickinsonia* was dwelling there were animals with bilateral symmetry, as well as heads. Working to decipher these fossils is not easy because, as they had not evolved teeth or bones, they didn't fossilize well.

'If I was working on dinosaurs, I'd find the bones, dig them up, take them to the lab and reconstruct the dinosaur,' Gehling says. 'But I'm dealing with soft-bodied creatures, and all you've got are imprints of squishy things living flat on the sea floor.'

Dr Jim Gehling examines the remains of what was once an ancient marine community in the Ediacara Hills in Australia.

# Ediacara

" There are certain places with names that resonate with historical importance. Places like Tintagel, the legendary castle of King Arthur, will be remembered for a very long time due to their importance. Ediacara is one such name; it's a place of unimaginable importance in palaeontology.

Most people don't realise that many of the geological periods were named after the place where rocks from that period were found. The Devonian period is named after the iconic red sandstone of Devon. Likewise, when the spectacular Precambrian fossils were discovered in the rocks of Ediacara, in Australia, the powers that be eventually deemed it suitable to name that whole geological period after the area.

What I find fascinating about the Ediacaran rocks is that they contain evidence of a community of creatures that lived just after fractal animals began to die out, yet these more recent creatures could do something totally new. About 560 million years ago, animals took a new and crucial step: they began to move. Although we aren't yet sure which were the first to do so, what we do know is that the animals that could move had a huge advantage over stationary rivals.

For me, Ediacara is a truly special place. It's in the middle of a dry and dusty wilderness, yet gazing at fossils encased in terracotta-coloured slabs of rock, I am transported to an ancient shallow sea teeming with life. The best time to search for fossils here is at first light, around 5.45 am. Not only do you avoid the searing heat of the afternoon, but you get low-level light, which highlights the fossils' edges brilliantly. If I'm honest, the thing that I like most about getting up early to hunt for fossils is that you can knock off around noon for a cold beer!

I first came here over 30 years ago to film some spectacular fossils of ancient jellyfish for the series *Life on Earth*. I've never seen those fossils since. Years after we filmed them, someone cut them out of the hillside, encased in 80 kg (176 lbs) of rock, and stole them. Eventually, the stolen fossils were traced to a warehouse in Tokyo. The Australian and Japanese governments became involved and only after nine years and a lot of negotiation did the fossils make their way back to a museum in Australia.

Fossils are so important to science and to our understanding of life on Earth. To have them stolen for the sake of money is a great and terrible shame."

*Spriggina* represents the first-ever animal with clear bilateral symmetry, where the left side of the animal is a mirror image of the right. And by having a head and a tail, this curious little creature is evidence of a major advance. Fractal creatures like *Charnia* relied on food coming to them, but *Spriggina*'s head demonstrates that some kind of sensory capacity had evolved – because there would only be a need to sense where food was if it had a mechanism to move towards that food.

*Spriggina* was first discovered in September 1957 by two private Australian collectors, Hal Mincham of Adelaide and Ben Flounders of Whyalla. They had taken the 400-mile trip north from Adelaide to follow in the footsteps of Dr Sprigg. They found numerous fossils, collected them and took pictures of their specimens. They passed the photographs to the South Australian Museum and the University of Adelaide for identification, where they were seen by palaeontologist Martin Glaessner.

Dr Glaessner immediately became interested because three of the specimens had unusual characteristics. To the untrained eye, they looked like the fossil of a fish, with a central spine and bones jutting out all the way down from the head to the tail. But Glaessner knew that most cnidarians do not have a head and that no other fossil of this age had ever shown any sign of such an organ. But all three specimens showed a bulge at one end that suggested some kind of head, a characteristic described as cephalization.

At Glaessner's request, the collectors donated their specimens to science and allowed them to be kept at the South Australian Museum for further study. In his first report to the scientific community, Glaessner described the three organisms as no more than 50 mm (2 in) in length, flat and segmented into around 40 sections, like a centipede with armour. However, if there were any hard parts associated with this animal in life, none remained in fossilization. Most intriguingly, he described how one side of the fossil was always larger than the other and almost horseshoe-shaped. This final characteristic hinted that they had cephalization.

Picking up from where Glaessner left off, Gehling has spent recent years considering *Spriggina*. He explains that while these animals look remarkably similar to the modern-day animal group of arthropods – animals like bumblebees, spiders, shrimps and lobsters that have jointed legs and segmented bodies – they were probably not closely related and also behaved rather differently.

'It's quite likely that *Spriggina* had sensory organs concentrated in the head,' Gehling says. 'Now why does my nose appear near my mouth? I want to smell the food before I ingest it. Why are my eyes above my mouth? So I can see what I'm eating. This head, seen in *Spriggina*, demonstrates that sensory capacity had evolved. This

*Spriggina*: this creature's bilateral symmetry is a major advance in evolution, and indicates that it was one of the first animals to move.

# A New Species?

"Every amateur fossil hunter has the same dream: to turn over a rock and discover a brand new species, previously unknown to humankind. Except that never really happens, or very rarely.

Well, I should count myself extremely lucky for getting anywhere close to that feeling. I was filming part of *First Life* with Jim Gehling at Nilpena, near Ediacara. He was working on some rocks and I happened to peer over his shoulder to see what he was doing. It just so happened that I had a better viewpoint than he did because of the direction the light was shining in, so I was the first to spot it – a new fossil. I showed Jim, who took a brush and cleaned it off.

He inspected it and said, 'Bless me, I can't match it to something I have seen before.' It's hard to explain the thrill of that moment of discovery. I suppose we never really lose that childish urge to explore and seek things out. I just love picking things up and looking at them. Knocking open a rock and knowing that you're the first person to see inside it, wondering what you're going to find, is a marvellous feeling. Finding something new is the icing on the cake.

We're still waiting to find out whether our discovery at Nilpena is a new species. There were a great number of *Dickinsonia* imprints on the same slab of rock, so it could just be a curious imprint of one of those. Jim Gehling has since found something similar in some rocks in Flinders Ranges, but his investigation is still underway. It would be terribly exciting if it were a new species, but either way, I'm just happy to be out there turning over rocks."

organism could sense where food was on the sea floor and so it is highly likely that it had a mechanism for moving towards that food.'

From more extensive research, we now know that *Spriggina* could reach 3 cm (1.2 in) in length and that its head was built from the first two segments of its body that overlapped.

On certain fossils, palaeontologists have noticed small depressions that may have been where eyes once existed. A few researchers have even suggested that the animals had antennae similar to those seen on modern-day insects. What is most fascinating about these tiny fossils is that on some specimens you can see something similar to a rounded mouth, but the fossils are so small that it is very hard to be certain about the details.

As for *Spriggina*'s behaviour, scientists do not know if they really crawled across the sea floor. Indeed, based upon their appearance, it looks as if they should have had jointed legs just as modern arthropods, such as insects, do.

Twenty years after the initial discovery of *Spriggina*, scientists were still puzzling over the find, tantalized by its arthropod classification. But, in 1976, Glaessner wrote a paper that outlined why he believed *Spriggina* was not an arthropod.

*Spriggina* failed the arthropod test on many counts. For starters, all arthropods have a specific number of segments on their heads, each bearing an appendage. *Spriggina* did not have the correct number of segments, nor was there any evidence of appendages. Furthermore, all arthropods have their mouths directed towards their posteriors (back towards their rear-ends), and *Spriggina* shows no evidence of this.

But if it was not an arthropod, what was it? Glaessner considered the possibility that it was an ancestor of the worms, which are collectively known as annelids. These animals today have numerous segments on their bodies but no legs. This similarity matched well with what was seen in *Spriggina*, although the fossil displayed a head that is far larger than that of any known worm.

*Spriggina* remains unclassified today; we do not know if it was a direct ancestor of a modern-day beast, or a species that became extinct somewhere along the evolutionary timeline. But the one thing that nobody contests is that it had a head, and for an animal that lived 550 million years ago, that opened exciting new avenues of exploration.

———

Even if *Spriggina* turns out to be closer to a garden worm than a butterfly, the fossilized tracks discovered by Martin Brasier and Alexander Liu at Mistaken Point prove that something alive could slither. The grooves of creeping animals have also been found in the younger deposits of the Ediacara Hills. It is now accepted that the simplest marine-dwelling worms came into existence between 565 million and 550 million years ago.

Worms may seem more complex than sponges, jellyfish or other cnidarians but, in terms of their basic structure, they are not. While cnidarians are effectively single sheets of cells that have come together to form a bag, worms are just a single sheet of cells rolled into a tube. Both groups of animals have insides and outsides, and they function in a similar way.

Cnidarians have surface cells and stinging cells facing outwards, as well as internal cells that secrete digestive compounds and absorb nutrients. Worms have a similar arrangement and, from an evolutionary perspective, these two creatures are close. It is easy to imagine a simple worm evolving from an early cnidarian – a hole in the centre of the bag and a little bit of stretching would do the trick.

In spite of their similarities, worms and cnidarians are behaviourally different. The jelly-bag cnidarians play the role of passive food gatherers. Those that live on the sea floor, like anemones, depend upon water to bring them food. Those that live in the water column, like jellyfish, do move around but still depend upon animals swimming into their long and nearly invisible strands of stinging cells.

Worms are more active, and movement is easier, so they seek food. The food goes in one end of the body and, after nutrients have been pulled out of the food, the excreted waste comes out of the other end. Our early rolled-up sheet of a worm was probably exactly that – a hollow tube – and it probably did not matter which way the food went in. Any nutrients that entered the tube were consumed, with leftover waste just drifting away. This inefficient set-up would have forced evolutionary pressures to select worms to use their tubes more efficiently.

The most likely scenario is that early worms quickly evolved digestive tracts that started with a mouth, had mechanisms to move food along the tube, and finished with an anus. Cells on the mouth would specialize in collecting food, cells in the middle would specialize in digesting, and cells at the end would expel waste material.

Sounds familiar? Even though the process appeared in worms so long ago, it is virtually the same system found in bodies of most animals alive today, including humans. We have mouths specialized with teeth for biting, tongues for food manipulation, stomachs that chemically and physically break apart food, intestines that absorb nutrients from broken-down food, and anuses that excrete waste. We have a lot in common with worms when it comes to basic digestion.

While the fossil evidence of a mouth and digestive system in the 550-million-year-old *Spriggina* is sketchy, there is fossil evidence of early digestive systems at the 530-million-year mark, though not in the traditional fossils that most would expect.

———

Dr Philip Donoghue at the Swiss Light Source in Switzerland uses the Synchrotron to discover the secrets of fossils that no palaeontologist

*Kimberella*: the earliest ancestor of the molluscs, this creature probably had a single muscular foot to pull itself along the sea floor.

could find with the naked eye. The Synchrotron is a giant microscope (housed in a building the size of a stadium) that reveals the tiniest detail, down to a thousandth of a millimetre. With it, Donoghue looks not just at the surface of fossils, as you would with an electron microscope, but actually inside the fossil using X-ray tomography (similar to a medical CAT scan), which allows him to build a 3D picture. It's this powerful, highly specialized microscope that allowed Donoghue to study fossilized embryos in the rock samples from southern China that we saw earlier in the book.

Now Donoghue is using the same technology to uncover the earliest digestive systems. Donoghue places a small rock pellet on the microscope stand. It is a 530-million-year-old fossilized egg, and he wants to see the embryo inside. As the high-energy X-rays of the Synchrotron bombard the fossilized embryo, Donoghue sits at his computer and zooms in on the specimen. At this scale, it looks monstrous, and as he focuses on one end of the tiny, worm-like animal, the image on the screen sharpens and we see a gaping mouth rimmed with terrifying spines.

Donoghue explains that the spines around the mouth are clearly developing into rings of teeth, and it is possible to see a developing gut that runs from the mouth to the anus at the end of the animal's tail.

'These fossils provide the first clear evidence for a gut,' Donoghue says. 'We can clearly see there is a mouth at one end surrounded by rings of spine-like body parts that would have played the part of early teeth. These extend inside the mouth where absorptive cells make up a gut that extends all the way through to an anus at the end.'

The animal, known as *Markuelia*, provides the first solid evidence of a fully formed digestive system. This was a key evolutionary innovation that allowed animals to actively seek food; in fact, it is the origin of the complex food chains found throughout the natural world today.

Seeking food was hardly an innovation 530 million years ago. Many species of single-celled organisms hunted for food in the Earth's early oceans long before *Markuelia* came along. Even *Dickinsonia* sought out bacterial mats on which to gorge.

What makes *Markuelia* special is that it is the first-known multicellular organism with the ability either to consume nutrients from sediments or consume other animals. The fossilized embryo of Donoghue's research would have hatched to produce a worm-like animal, just a few millimetres long at first and with many segments from head to tail. Upon reaching adulthood, it would have grown into an organism that was tens of centimetres in length, complete with a fully developed digestive system.

Rather remarkably, there are animals today that fit this description, such as *Priapulida*, a type of marine-dwelling worm. These worms, popularly known as penis worms because of their shape, have bodies

Philip Donoghue's scans of *Markuelia* from the Synchotron show the first sign of a digestive system, and a ring of teeth suggesting that it was a predator.

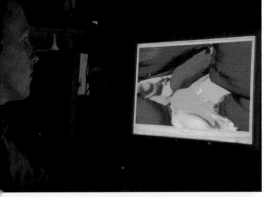

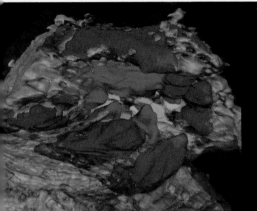

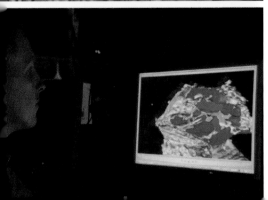

that are strikingly similar to *Markuelia*. They spend most of their lives burrowing through sediment. They actively seek food and extend a tube from their mouths, which are surrounded by rings and rows of teeth. These teeth are used to grasp hold of living and dead organisms, and drag them into their mouths. Around 530 million years ago, you would probably find *Markuelia* doing the same thing – poking their heads out of the mud on the sea floor and, perhaps occasionally, pushing out their spiky mouths to drag a tasty morsel into their burrows.

Donoghue is uncertain whether *Markuelia* was hunting or scavenging, but its resemblance to *Priapulida* makes it likely that it was at least burrowing through sediment and seeking food. 'The fact that it has teeth around its mouth protruding outwards means that this thing was a predator,' Donoghue says.

———

The burrowing we have seen in ancient worms at Mistaken Point implies that these worms could move. The evolutionary benefits of movement are quite clear, but how the earliest animals actually managed to move from point A to point B is still not known.

Robert Clark at the University of Newcastle proposed a theory in 1964 that has appealed for decades. Most worms today have an internal cavity in their bodies called a coelom. The cavity is filled with fluid and helps to give the worm both a rounded form and structure. Dr Clark theorized that if an early worm had muscular cells that could squeeze the fluid-filled coelom, this would have caused one end of the tube to extend itself a bit, a tiny push that might have been the early mechanism of movement.

To visualize the primitive worm wriggle, you need to find the kind of long balloon that children's entertainers twist into the shape of a dog. If you half-fill this balloon with water, you should see that part of the balloon expand to its full diameter, while the other part of the balloon would remain unexpanded. If you squeeze the full section of the balloon with your hand, the water squirts through your grip, travels to a different section of the balloon and inflates to its full diameter. The section in your fist is now all but empty.

The worm coelom was similar to the half-filled balloon. If the worm pushed particularly hard with its muscles on one section of the coelom, this would cause fluid displacement and expansion of the coelom at a different part of the worm's body.

With the balloons, it is possible with a little twist or stretch of the rubber to expand only a small section of the balloon. This is possible because stretching a specific section of balloon weakens the rubber and, as air or water is blown in, that weaker section expands first. Following from this concept, if one section of the worm's coelom cavity wall was slightly weaker than other sections, it would be this section that filled with fluid first.

So, in principle, if you had a long balloon that was stretched a bit at one end before being half-filled with water, that end would fill up first, and as the water pressure from filling increased, the areas near that stretched section would fill as well. Then, if you squeezed the filled end in a hand-over-hand method from front to back, the fluid inside would slowly be pushed to the far end of the balloon. Once at the far end of the balloon, if you were to continue to try to compress the water, with nowhere to go and pressure from your relentless squeezing increasing, the weak section at the front of the balloon would suddenly be injected with water from the back and expand. Now imagine that the balloon was a worm body. If this weaker section of the worm were inserted into a bit of sediment, the sudden expansion of it by fluid compressed by muscle activity would give the worm an anchor point, and as the worm further inflated the front of its coelom, this would allow it to pull itself forward into the sediment by systematically contracting its muscles.

Clark's theory on worm movement did not strike him out of the blue. While the worms found in today's gardens do not use muscular contraction of their coelom to move around, there are some worms that contract their muscles around fluid-filled cavities to extend parts of their bodies. Some use these methods to extend their mouths to capture prey; others use the pressure of fluid-filled bags to extend reproductive organs for the transfer of eggs and sperm. The existence of such mechanisms today does not prove that worms first started to wriggle in this fashion, but it does suggest that such mechanisms do readily evolve in worms – and may have been the first form of propulsion.

Regardless of how it evolved, the ability to burrow would have granted early worms a tremendous boon. Until the evolution of burrowing animals, any dead or decaying material that fell to the ocean floor that was not consumed immediately would have been buried by sediment – where it would rot away or eventually be fossilized. Fossilization might be great for palaeontologists today, but it was an enormous waste of food for hungry animals. By burrowing, worms could suddenly exploit this source of buried and untapped food.

But burrowing would have introduced a new problem: burrows are not good for breathing. Those first burrowers like *Markuelia* could not stay under the mud for long.

We all need oxygen or we suffocate, and as mammals we can only get it from the air. If we dive into the sea, we can stay beneath the surface for only a couple of minutes, holding our breath. Even animals that naturally live in the water need this vital gas. Fish, clams and ocean-dwelling worms have mechanisms that allow them to collect oxygen dissolved in water. If oxygen were to be removed from the water, these animals would suffocate. And this is exactly the problem that burrowing worms run into: oxygen in sediment is nowhere near as rich as it is in most sea water. As those first worms went digging for food, they would not have been able to go down very deep or for very long.

Although the size and shape of early worms were beneficial for movement, to make matters tougher, the worms would have been struggling to get oxygen to all their cells. Oxygen enters the cells through a physical process known as diffusion, where it moves from high concentrations in the water, across the cell membrane, into the oxygen-starved cell. The same is true with the waste gas carbon dioxide. If a cell has a build-up of carbon dioxide and is exposed to water with a lower concentration, then the gas will naturally, through diffusion, travel out of the cell and into the surrounding water.

This process of collecting oxygen and expelling carbon dioxide keeps 99.9 per cent of all organisms alive. But if an organism absorbs oxygen through passive diffusion, each cell needs to be exposed to oxygen-rich sea water.

—

*Fossilization might be great for palaeontologists today, but it was an enormous waste of food for hungry animals. By burrowing, worms could suddenly exploit this source of buried and untapped food.*

For fractal frond-like organisms, acquiring oxygen was easy. The fronds were wide and thin, and all the cells were exposed to water as it flowed by. Sponges are similar – filled with thousands of holes that let water, with dissolved oxygen, flow through them. Even cnidarians manage oxygen collection in this simple way, regularly flushing water and dissolved gases in and out of their bag-like bodies.

By being long and tube-like in shape, early worms immediately had a major flaw: they could not easily deliver oxygen to some of their cells. As oxygen with water entered the mouths of early worms, cells at the front of the long, tubular animals would have easily collected the required oxygen from the water and subsequently released carbon dioxide through diffusion. Further along their bodies, more cells would collect oxygen and release carbon dioxide. This gas-exchanging process would alter the chemistry of the incoming water as it travelled through the body, the result being that the middle part of the early worm's gut would have been depleted of oxygen but rich in carbon dioxide. Diffusion of oxygen into cells could not take place, and the cells would surely die.

But long worms thrived in this period. Scientists believe their solution to oxygen starvation was segmentation. Segmented bodies

# The End of Prehistoric Fractal Fronds

Bilateral bodies are on the rise: guts, muscles, circulatory systems and hard parts have become the latest must-have accessories. These evolutionary additions to multicellular organisms also meant that animals were more actively involved with their environment. But those strangely beautiful fractal fronds, bending gently in the ocean currents, began to decline.

Guy Narbonne suspects that the simplicity in their design is what allowed fractal organisms to get a head start on many other multicellular life forms and dominate the planet so quickly. Creatures like *Charnia* probably had a simple genetic code controlling body formation, as fractal design does not require much information for body development. Despite their size, the fractal organisms were genetically basic; in fact, Narbonne believes they could be put together with between six and eight 'genetic commands'. This stands in stark contrast to the 25,000 'genetic commands' required to construct a complex mammal like a human.

However, Dr Narbonne also suspects that this fast start was ultimately what led to the fractal organisms' decline. Their simple evolutionary design probably left them unable to evolve complex structures like teeth, claws, muscles and brains. Without these features, he theorizes that they started to dwindle away, while animals with the ability to move considerable distances and animals with hard parts in their bodies started to appear.

Animals built from a fractal blueprint were like sprinters bursting from the starting blocks of evolution. They led for the first 15 to 20 million years of multicellular life on the planet but, as the race progressed, they fell out of the race as animals with more complex body forms overtook them.

**Guy Narbonne believes that it was the simplicity of fractal fronds that ultimately led to their decline.**

are common on Earth today, found in caterpillars, honey bees, scorpions, lobsters, shrimps and thousands of other animals. Worms are also segmented: the segments split their coelom into individual chambers that are connected to one another – and, crucially, also to the skin of the worm – through a series of valves.

With this set-up, water with dissolved gases does not merely get pushed through the body from head to tail and absorbed on a first come, first served basis. Instead, it also enters small holes on the worm's skin and is pumped by muscles around the body before it is exposed to cells that need to exchange gases. This shares the oxygen fairly, giving every cell access to oxygen-rich water.

We have already seen that people and worms have a lot in common, and here is another shared feature: the system in our own bodies for getting oxygen to every cell in the body is not that different from the worm's water-pumping mechanism. We breathe air that is rich in oxygen; it is drawn into our lungs and then transferred into the blood. The heart pumps this oxygen-rich blood to arteries that quickly branch off in different directions and each carry off equal amounts of the vital dissolved gas. As the arteries branch, they become smaller and smaller until they are exceedingly thin and pass directly through body tissue made up of cells that are required to collect oxygen and release carbon dioxide. Our clever system of blood vessels allows this gas exchange to take place through diffusion at points of the body a long way from where the oxygen first enters it, and, in turn, carbon dioxide is carried off and released from our bodies when we exhale.

—

*By analysing human embryos as they develop, and by comparing them to the embryos of worms, it is obvious that worms are our ancient ancestors. We are not just like worms; we are descended from them.*

This similarity is no accident. By analysing human embryos as they develop, and by comparing them to the embryos of worms, it is obvious that worms are our ancient ancestors. We are not just like worms; we are descended from them. The coelom that they evolved developed in humans and most mammals into the central cavity of our body within which all of our major organs today are held. And the worm mechanism for collecting oxygen and releasing carbon dioxide is what ultimately evolved into the respiratory and circulatory systems found in insects, fish, amphibians, reptiles and mammals.

But what about our suffocating worm hiding in its muddy burrow? While the water-pumping mechanism helped worms to manage the problems created by their own tube-like shape and larger bodies, it did little to help them cope with the oxygen deprivation in the slime of the ocean floor.

Once again, we can turn to modern worms for the answer. They deal with the oxygen depletion in sediment in a number of different ways. Some worms have long strands of tissue that they send out from their burrows that function a bit like a snorkel. They are connected to the circulatory system, draw in oxygen, expel carbon dioxide and allow a worm that is buried well beneath the surface to allow its cells to exchange gases.

What is interesting about this evolutionary approach is that is not a major leap for that strand of tissue absorbing oxygen to evolve characteristics that might allow it to collect food as well, and this is common among modern worm species. Many modern worm species use these strands to filter particles of nutrients out of the water. On some worms, the strands have developed into tentacles that allow the worm to grapple and kill prey. Indeed, with the bulk of the worm's body buried beneath the sediment, it is difficult for animals swimming or crawling past to foresee their fate. They are caught by surprise, and the worm settles down to eat.

———

But the tactic of leaving strands of tissue hanging out of the burrow for respiration creates a major limitation for the worms that use it. Like snorkellers who cannot dive deep, early burrowing worms using breathing tubes would have had to stay in shallow sediment. In addition, worms that use the snorkel strategy are not free to burrow and breathe at the same time. If they want to move, they must either reel in their breathing apparatus and cope with poor access to oxygen while they travel, or leave their burrow and travel above the thick sludge.

A different method used by modern worms to get oxygen to their tissues while underground involves them moving their bodies rhythmically as they burrow. The rhythm functions like a pump and draws water from the surface down into the burrow. As this fresh, oxygenated water reaches the worm, the water around the worm, which is rich in carbon dioxide, is displaced and pushed out of the burrow. Using this method, some worm species can burrow and breathe at the same time.

———

Worms were clearly innovative creatures, quickly evolving features to solve the problems that their tube-like shape presented. But the fossil record reveals that these primal wrigglers were not the only animals moving about the ocean floor. There are dozens of other markings that suggest movement but do not look like the tracks of worms. In some cases, there are even fossilized animals near these markings. Could these be the creatures that made the tracks?

One such fossilized animal from sediments that were roughly 555 million years old in the Ediacara Hills came to Dr Glaessner's attention in the early 1960s. The fossil had an elongated body, oval in shape and tapered at one end, and it appeared to be a cnidarian. Glaessner classified it as such in 1966 and named it *Kimberella* after a devoted fossil collector named John Kimber who had recently died.

More than 30 years after the discovery of *Kimberella*, palaeontologists re-examined its place in the animal story. During this three-decade-long period, more fossil beds of a similar age containing specimens of *Kimberella* had been found in Russia near the White Sea, alongside scrape marks and meandering trails in the sediment nearby. Mikhail Fedonkin, a palaeontologist at the Russian Academy of Sciences, spent a lot of time examining these fossils, picking up where Glaessner left off. But Dr Fedonkin argued that Glaessner had missed a critical feature. Instead of having radial symmetry, as all cnidarians possess, *Kimberella* appeared to be bilateral. Moreover, the scratch marks near *Kimberella* fossils resembled the marks that are often associated with modern snails.

Snails and slugs may look soft and squishy, but many gardeners know all too well that inside their bodies are the tools that can shred plants to pieces in a single night. The part of a snail's or slug's anatomy that allows them to cause such damage is a ribbon-like tongue called a radula. Like a double-edged saw, it is covered in tiny sharp teeth.

Slugs and snails are more vicious than most gardeners appreciate. Today, there are a few carnivorous slugs that actually use their radula to kill earthworms, while another family of sea snails uses its radula as a weapon to harpoon nearby prey.

Radula do not fossilize well. Even when palaeontologists look at recently fossilized snails and slugs, they rarely find the radula preserved. However, the scrapes that snails and slugs leave behind have a tendency to fossilize very well, and these trace fossils were what inspired Fedonkin when he was looking at the marks near *Kimberella*.

Based upon the presence of bilateral symmetry, a trait that snails and slugs share, and the likelihood that the scratch marks were created by a snail-like radula, Fedonkin proposed in 1997 that the *Kimberella* was more likely to be a member of the group that contains snails and slugs, the molluscs.

———

Whether *Kimberella* was a mollusc or not, the radula-like scrapings near these fossils and its meandering trails show that *Kimberella* was moving. Yet it was clearly not a worm. Interestingly, in the years after Fedonkin's 1997 paper, some fossils of the animal were found to contain another fascinating detail: a bit of semi-hardened shell. The shells that Fedonkin and his colleagues analysed were no more than 15 cm (6 in) long and 8 cm (3 in) wide, but their presence made palaeontologists wonder: what could a piece of shell possibly have been used for?

For now, the answers rely on scientific speculation rather than conclusive evidence, but theories have developed based on observations of the modern and prehistoric worlds.

One idea formulated by scientists was that this early shell functioned as a place to which the muscles in *Kimberella*'s body could attach. However, hard parts are not a prerequisite for muscles to evolve. We have seen animals such as the *Dickinsonia* disc and the worm *Markuelia* both functioning perfectly well without anything hard holding them together. Even if an animal is soft and squidgy, it can still withstand contractions, as long as the soft tissue to which muscles are attached is strong enough to withstand the force.

This has caused many palaeontologists to conclude that *Kimberella* was using its shell for another purpose: respiration. Many organisms maximize the amount of gas they exchange with the surrounding water or air with structures that have large surface areas. Lungs follow this principle. They have thousands of branching pathways through which inhaled air can travel; the surfaces of all these pathways and the tiny air sacs at the end of every branch can readily absorb oxygen and release carbon dioxide. These surfaces add up to a very large effective surface area – equivalent to that of a tennis court for each pair of human lungs – which dramatically improves gas exchange.

Some animals have surfaces outside of their bodies that help them perform gas exchange. Amphibians, such as frogs and salamanders, exchange gas through their skin. In this case, their entire bodies play a role that is much like that of the inside of lungs and help them to respire. This causes palaeontologists to wonder whether *Kimberella*'s enigmatic shell might have had a thin layer of tissue on top to exchange gases.

Fossils from the Ediacara Hills.

———

The specialization of animals made possible as a result of bilateral body plans took place extremely quickly in the evolutionary timeline. It happened over just a few million years, rather than the billions of years it took for animal life to appear in the first place. In a relative geological instant, the gene pool became extremely varied and diverse. There is only one explanation for this: sexual reproduction had evolved in animals.

In the last few years, the Ediacara Hills have given up one of their most remarkable secrets with the discovery by American palaeontologist Mary Droser of an animal called *Funisia dorothea*. This creature lived in colonies, with each organism anchored to the sea floor, wafting around in the ocean currents. Dr Droser, based at the University of California, Riverside, and Jim Gehling believe this is the first evidence of an animal with a sex life.

Amid all of the physical innovations in the fossils of the post-Snowball Earth world, it is easy to miss *Funisia dorothea*. It is just a long, thin, worm-like organism that once stood upright, shin-high from the sea floor. Researchers have not even been able to identify a recognizable anatomy.

Of course, for life on Earth to evolve to a point where animals like *Dickinsonia*, *Spriggina* and *Charnia* could exist on the planet in just over a couple of billion years, sexual reproduction had to have existed for a long time. As we saw in Chapter 2, it is the only way to explain the rapid pace of change and evolution. But scientific discovery doesn't follow the evolutionary timeline, and Droser's find is the earliest actual evidence of sexual reproduction identified in the fossil record.

When animals clone themselves, and mutation is the only mechanism for change, evolution takes a long time because there is only gradual change to the genetic blueprint. But when bacteria swap genetic material among themselves (a process known as conjugation), or when higher organisms combine their genetic material to create unique offspring, evolution is much faster. The complete shuffling of genetic information of two individuals creates amazing variety and the speed of evolution rockets ahead. For life to have jumped from simple single-celled organisms to creatures as complex as those found slithering around in the post-Snowball Earth world, sex had to have been present.

'Sexual reproduction is absolutely one of the most fundamental steps in the history of life on this planet,' Droser says. 'It is why we have the diversity that we have. As far as we know, this is the first evidence of sexual reproduction in animals. We're not catching the animal in the act of it, we're looking at the product of what we conclude was sexual reproduction.'

The trouble is, sexual reproduction as an activity does not tend to fossilize well. Yet Droser suspects that she is on to something with *Funisia dorothea*. Fossils of the column-like organism are almost always found in groups, like small fields of Brussels sprout plants on the seabed. More importantly, these groups are always composed of organisms that are of roughly the same size, and their stems all have the same diameter. This suggests, according to Droser, that these groups were all 'born' at the same time, grew up together and are genetically similar.

Near these groups there are also often patches of little pimple-like structures that fossilized. Droser found these pimples were in a transition stage to becoming adult *Funisia dorothea*. The patches were areas where young versions of the organisms were settling and getting ready to grow.

None of the fossil evidence catches *Funisia dorothea* in the act of having sex, but the way its offspring are growing suggests that sex was taking place. Droser is confident in her conclusion because when you look at modern environments where these sorts of size and age groupings occur, sexual reproduction is almost always the cause. As is often the case, the past is studied through the present.

Droser draws most of her evidence from corals. Corals of the same species in a region release all of their eggs and sperm at exactly the same moment once a year. By releasing such vast quantities of sex cells, called gametes, all at once, the corals give their sexual cells the highest chances of bumping into one another, fertilizing the eggs and forming embryos. These embryos settle at the bottom of the ocean in groups at roughly the same time and begin to grow. So the animals in the newly formed groups in a single area are typically similar in size and shape because they were all fertilized simultaneously and probably came from the same parent corals. The clustered beds of *Funisia dorothea*, where all the fossils are the same size, show for the first time in the fossil record indirect evidence of sexual activity.

Droser is the first to admit that the evidence suggests, rather than proves, that there was sexual activity among these organisms. She also argues that there were probably many other organisms alive that engaged in sexual reproduction. However, because there is no evidence of what exactly they were doing, it is impossible to argue the matter convincingly.

*Spriggina*: a head indicates that some kind of sensory capacity had evolved, allowing this creature to sense where food was on the sea floor and move towards it.

# Funisia and Sex

It's not a question that occurs to most people, but why bother with sex at all? Sex is a very costly process, both in terms of time and energy. You only have to look at the remarkable bowerbirds to question the folly of sex. The males of this utterly charming bird family spend many days constructing nest-like structures, called bowers, out of twigs. Content with the shape, they set about decorating them with pebbles, dead insects, leaves, even human rubbish in some cases. These intricate bowers might look like nests, but they will never contain eggs. Their purpose is entirely ornamental.

This curious ritual is a wonder to observe, but for the male bowerbird it takes up an enormous amount of time and energy, both of which could be well spent feeding or looking out for danger. Why do the males bother? They do it in the hope of attracting and impressing a female bowerbird. If a female finds a male's bower appealing to the eye, she will mate with him, and bear his young. The whole rigmarole boils down to one issue: sex.

For animals to bother with sex at all it must have some pretty strong evolutionary benefits. These benefits are actually wonderfully simple. The act of mixing your genes with another creature allows you to create offspring which are genetically diverse, and more likely to survive in a changeable environment.

When Mary Droser discovered fossils of *Funisia*, she noticed that they were not randomly distributed across the rocks on which she found them, but gathered into groups. What's extremely fascinating about these grouped fossils is that although individual fossils in groups were of the same size, their sizes varied overall between groups.

Crucially, this meant that *Funisia* was being distributed to different points, from which individuals of the same size and age grew. This was the first evidence of sex.

It is quite unlikely that *Funisia* was the first animal to have come across sexual reproduction, but actually, what is more thrilling is that it suggests that animals in the Ediacaran period were reproducing by sexual means. If this process of mixing genes was being used more widely, it could perhaps explain why animals at this time suddenly increased their level of complexity. Sex effectively sped up the rate of evolution, producing genetic variation at a far greater speed than ever before. Fossils of *Funisia* might not be terribly beguiling to look at, but they could represent the first great sexual revolution, and that's pretty exciting.

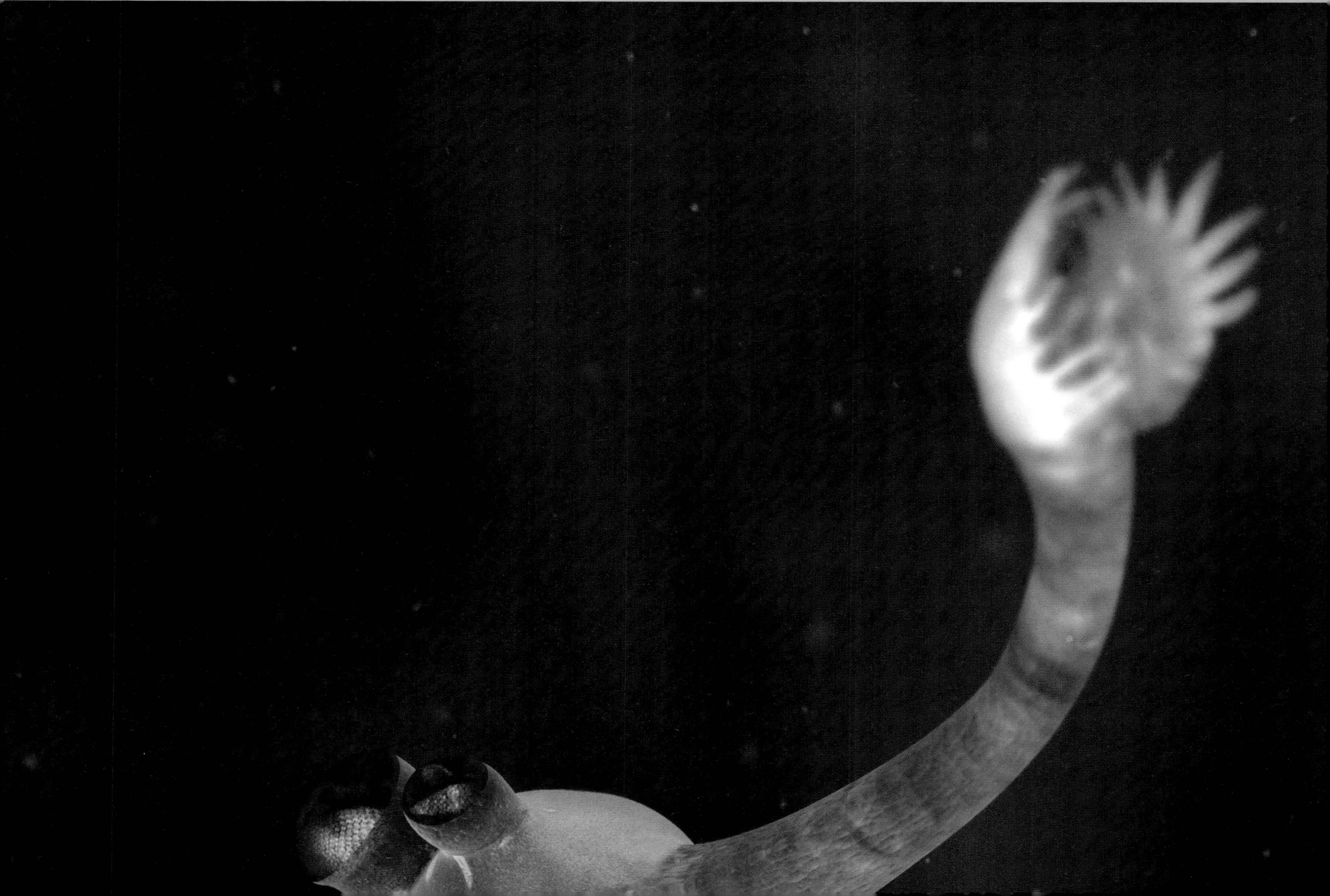

IT IS LATE August 1909, high in the Canadian Rockies. Charles Doolittle Walcott, an American invertebrate palaeontologist, picks his way across a narrow path that traverses a vast, steep tumble of scree. This is his second trip to the mountains, lured by reports of fossils being found by workers on the Canadian Pacific Railway.

He pauses to enjoy the afternoon sun and when he turns back to the path and looks down to find his footing, he sees the most beautiful fossil, preserved in astonishing detail and like no other he has ever seen before.

This is how Charles Walcott discovered the Burgess Shale, a band of mudstone and shale that is stuffed full of fossils – some of the most extraordinary creatures that the world has ever seen. This find gave palaeontologists an unprecedented porthole into life in the seas around 505 million years ago in the Cambrian period.

Scientists have since excavated more than 65,000 different specimens of extinct animals from a small quarry in Burgess Shale, now known as Walcott Quarry. Many of these species have never been found anywhere else, but what's more extraordinary is that the soft body parts are preserved. These fossils have transformed our understanding of the evolution of modern animals. While many of them are of animals so bizarre that they are unrecognizable, others are the first glimpses of animal shapes and forms that are somewhat familiar.

Before we prise open the shale and sift through the finds from this amazing site, we must first look at the chronology of events – not just the geological chronology but also the timeline of fossil discoveries.

Thus far, we have explored the Ediacaran period, that span of time from the final thaw of Snowball Earth around 630 million years ago to 543 million years ago. This significant period marks the beginning of the Cambrian period, when life in the seas seems to have increased at an exponential rate and in the most dramatic way. Thanks to discoveries from the Burgess Shale and older Cambrian fossil sites, we see that over a 10- to 20-million-year period – the snap of a finger in geological terms – life evolved and diversified at a breakneck pace, resulting in creatures that could hunt and hide, see, chew and even walk. Early palaeontologists called this the Cambrian Explosion.

The Burgess Shale in the Canadian Rockies has yielded some of the most extraordinary fossils from the Cambrian period found anywhere in the world.

We know now that the origins of these forms were already flourishing long before the start of the Cambrian period 542 million years ago. We have seen, for example, the teardrop-shaped *Spriggina* with its distinct head, the bulbous burrowing worm *Markuelia* with a mouth and teeth, and the enigmatic *Kimberella*, perhaps the first-ever animal with a hard shell. But in the first 10 to 20 million years of the Cambrian period, we find evidence in the fossil record of an extraordinary growth in the diversity of life.

So while the pace of evolution at the start of the Cambrian period is remarkable – and the fossil record shows a significant increase in the number, diversity, complexity and size of animals as never before – we can see that evolutionary pressures were already driving organisms to experiment and specialize. But this is *not* how it would have looked to palaeontologists in the early 20th century.

Our insight into the seas of the post-Snowball Earth comes from a few rare fossils beds, especially those in the Ediacara Hills of the arid Flinders Ranges of South Australia. However, the fossils locked in these old black rock formations were not discovered until half a century after Dr Walcott's serendipitous discovery of the Burgess Shale. Up until the 1950s, palaeontologists generally believed that complex life first appeared 543 million years ago. To them, the situation was simple; in rock from the Cambrian period and later, layers of fossilized shells, often referred to as 'small shellies', were present. In older rocks, small shellies could not be found. Life clearly looked to have simply

# Opabinia

" Whilst filming *First Life* I was able to retrace the steps of Charles Doolittle Walcott through the Burgess Shale site in the Canadian Rockies and imagine his reaction to finding, lying on the shale, a beautiful, tiny fossil of a kind he had never seen before: *Opabinia*.

No creature like *Opabinia* exists on Earth today. It was very much an evolutionary experiment – a bizarre animal with five mushroom-like eyes. There are clues as to how this creature may have lived: it lacked legs but had a broad tail and flaps along both sides of its body. Computer reconstructions of the fossil suggest it moved by wafting these flaps, giving it great flexibility of movement in the water.

*Opabinia* also possessed a flexible proboscis on its head with which it grabbed food from the floor of the shallow sea in which it lived. It was a truly primitive creature and one that left no descendants. Walcott was wise enough not to try to classify *Opabinia* amongst either the annelid

worms or the arthropods that arose around that time, and scientists are still somewhat baffled as to where it fits in the evolutionary tree of life.

*Opabinia* wasn't alone in the Cambrian seas. Countless other bizarre creatures burst onto the scene at the same time. It was an unprecedented surge of diversity, something that had never happened before and has not happened since.

The creatures found in the Burgess Shale fascinated Walcott, who returned there many times over the 15 years following his initial discoveries. He even brought his family along to help, and together they amassed an astonishing 65,000 fossils.

Despite the Walcotts' huge contribution to our understanding of the Cambrian period, and the work done by palaeontologists since then, there are still many unsolved questions about these perplexing creatures, questions that I'm not sure we'll ever manage to answer."

come out of nowhere. From what they could see, life in rocks that were 543 million years old and younger was both extremely common and exceedingly varied. But there was no build-up to life's sudden diversification, just a sudden great explosion of diversity.

The belief that life suddenly and rapidly diversified at such a specific point in time created real problems for Charles Darwin. Darwin is famous, along with the lesser-known Alfred Russel Wallace, for proposing that animals evolve as a result of natural selection. Darwin concluded this after studying the natural world, both at home in England and while travelling the globe on HMS *Beagle*.

In December 1831, the *Beagle* left Plymouth, heading southwest to explore and survey the coast of South America. On board was the young Darwin, officially the ship's naturalist and geologist, but really there to provide company and conversation for Captain Robert FitzRoy. FitzRoy had feared the long periods of loneliness during a five-year circumnavigation, with only the creaking masts, flapping sails, howling wind and the rough company of sailors.

Darwin's observations of the animals and plants that he encountered during the voyage were seminal to his theory of evolution. He saw fantastic creatures in a variety of environments, ranging from pigeons with bizarre feather formations and fossilized shelled animals in the mountains of South America to giant tortoises and saltwater iguanas. In particular, he noticed the effect that islands seemed to have on life. He documented the many different finches on the Galapagos Islands in the Pacific, and speculated that habitat and food availability drove them to become different.

Fossils of *Anomalocaris* found at the Burgess Shale, showing the claw (top) and complete body plan (bottom).

---

The lesser-known Alfred Russel Wallace was also a naturalist, and had worked as a civil engineer, surveyor and teacher. He delighted in the outdoors and shared Darwin's passion for collecting beetles. An avid reader of Darwin and other great naturalist explorers of his time, Wallace decided to go on adventures of his own, the first being to the Amazon. He lost most of the specimens he had collected in a fire on board during the sail back to England.

At 31, Wallace embarked on an eight-year expedition through the East Indies (now Malaysia and Indonesia). In Indonesia, he noted clear biological differences between animals on neighbouring islands. He realized that the animals on some islands in the region were closely related to species more commonly found in mainland Asia. Animals on other islands were closer in relation to species in Australia. Overlap among species from the different regions of the world was almost nonexistent. These discoveries left him thinking about the isolating effects of islands and how isolation might drive animals to evolve.

Even though Darwin and Wallace independently concluded that evolution via natural selection was likely responsible for the incredible diversity that they were seeing in the living world, doubts remained. Darwin showed that he was truly a great scientist by being remarkably willing to criticize and attack his own arguments. In his landmark book explaining evolutionary theory, *On the Origin of Species,* he wrote at length about the serious questions presented by the sudden appearance of life some 543 million years ago. He felt that if life gradually changed over time, as natural selection suggests it should have, it would make sense to see a steadily increasing diversity of animals in the fossil record. The apparent absence of fossils from earlier than 543 million years ago made no sense, and he openly suggested that this evidence stood strongly against his own arguments.

But the 1909 discovery of the Burgess Shale provided an explanation for Darwin's dilemma. Walcott returned to the site with his sons the following year and, over the next 14 years, he carried out extensive excavations, finding thousands of specimens in the mudstone layers of Walcott Quarry.

Unlike the bits of shell and broken body armour that were common in the small shellies in Cambrian rock, Walcott found impression fossils that often revealed the bodies of animals with soft body parts. These were preserved as thin, almost imperceptible layers, noticeable only if you get the light just right.

Walcott theorized that the animals were the ancient relatives of many modern animal groups and suggested that they lived in a steeply sloping, muddy environment below an oceanic reef. Every now and then, landslides of fine mud would fall catastrophically down the slope, taking with them a host of animals that would be buried alive when the mud settled at the bottom. Then, over time, the mud transformed into shale and was uplifted by the same geological forces that have raised the mighty Rockies along the entire length of North America. The perfectly preserved Cambrian creatures eventually found themselves locked in stone near the summit of Canada's highest peaks.

Immediately, Walcott realized that the Burgess Shale fossils hinted that early palaeontologists had been wrong: life had not jumped into existence; it had been present all along but was effectively invisible.

What is most remarkable about the enormous Burgess Shale fossil collection is that it contains a wide diversity of animals that had no hard parts at all. The unique geology of the Burgess Shale made it possible for Walcott and his fellow palaeontologists to see the diversity of life that had been present 505 million years ago. Up until this discovery, the fossil record had been biased - fossilization is much easier for hard body parts, so their fossils are much more commonplace. The particular conditions that created the Burgess Shale - a sudden fall of fine mud burying organisms in a low-oxygen environment - preserved animals that would otherwise have rotted away, leaving no trace of their existence.

# Collecting and Classifying

❝ Although fossils were my main passion as a boy, I loved collecting other things as well: birds' nests, rocks and snake skins. I collected just about anything that gave me a sense of the world around me, of its diversity and history.

When I was around 16 years old, I planned a very ambitious trip in order to understand the geology of the Lake District. I made myself a couple of canvas bags to fill with geological specimens, and a long box, which I filled with straw. I fitted the bags to my bicycle and posted the box to the goods department of a station along my route.

Once I'd collected several hundred little specimens I was able to put them in the box when I caught up with it, and send it on to the next station. This way I managed to do a whole circuit of the Lake District collecting rocks and fossils, staying in youth hostels along the way.

It was a marvellous trip and I collected a whole pile of specimens for my collection. In the end I had so many bits and bobs that my father allowed me to create a little museum in the University College of Leicester, where he was principal.

I'm very glad I developed my love of collecting; it's a very valuable way to engage in natural history and it taught me a great deal. By collecting objects and examining them you can work out a system of classification and notice variations. It allows you to understand which species or forms are common and which are rare.

Classification and identification form the basis of natural history. It's no accident that Charles Darwin, arguably the greatest natural historian who ever lived, was crazy about collecting beetles. He was determined to discover a new species before his beetle-collecting rival, Charles 'Beetle' Babington. He once spotted a new beetle and as he already had a beetle in each hand, he put one in his mouth to keep it captive while grabbing the new one. Unfortunately, this cantankerous beetle fired acid into his mouth, causing him to drop and lose the new species!

I'm not sure whether he managed to beat his rival, but over those years of collecting he most certainly developed a keen eye for fine detail. This undoubtedly helped him to spot the similarities and differences between the famous finches in the collections he brought back from the Galapagos Islands.

I could never compare myself to the great and inspired Darwin; I'm just a broadcaster with a passion for the natural world. I would say, however, that my love of collecting certainly has helped me to understand and greatly appreciate the sheer variety of life that exists around us, and indeed the variety of life that came before us." ❞

Walcott concluded that the absence of fossils from before 543 million years ago was not evidence of the absence of life. Instead, he decided that life was probably quite common, but it simply had not fossilized very well because it lacked the hard body parts advantageous for forming fossils. Darwin need not have worried, it seemed, because the apparent Cambrian Explosion was merely an artefact of fickle fossilization, marking the evolution of hard body parts.

Throughout the 20th century, the idea of a Cambrian Explosion of life became less and less convincing as more impression fossils of soft-bodied organisms were found at sites of fine-sediment Precambrian rocks. Indeed, the rich fossil beds in the Flinders Ranges in South Australia and Mistaken Point in Canada revealed the great diversity of life in Precambrian times, with seas that had obviously teemed with a wide variety of life.

However, in the 1970s, two palaeontologists, Niles Eldredge at the American Museum of Natural History in New York and Stephen Jay Gould at Harvard University, proposed an idea that would once again support the concept of a Cambrian Explosion, while simultaneously showing that a sudden diversification in organisms 543 million years ago need not shatter the theory of evolution via natural selection. They simply proposed that evolution at this point went into hyper-drive. But why?

**With Dr Jean-Bernard Caron, the world's leading expert on the Burgess Shale.**

Darwin and Wallace had both theorized that life would gradually change over time. They argued that when species encounter new challenges in their environment, the variations that naturally arise in the species' population through mutation and sexual reproduction allow the species to generate new traits. These variations would allow some individuals to handle changes or challenges better in their environment, and they would breed more often than their kin as a result, ultimately spreading their novel traits (just as we saw in Chapter 3 with the theoretical example of the castaway blue parrots).

The idea of evolution via natural selection made perfect sense when Darwin and Wallace suggested it and, more importantly, biologists studying the modern natural world regularly see this evolutionary theory demonstrated with many different species.

If you want to see evolution in action today, just visit a farmer's field close to home or, if you feel a little more adventurous, venture into the malarial swamps in Africa. Both locations are likely to have been treated with pesticides, the farmer trying to kill off crop-destroying beetles, and public health authorities battling against disease-carrying mosquitos. Spraying insects with toxic chemicals helps to control pest populations, at least initially. Almost all of the pest insects in an area die out from the toxin, so that crops thrive and the spread of malaria is reduced.

The pesticides, in other words, pose an environmental challenge to these insects, but experience shows that the populations of insects quickly evolve. Unfortunately for the humans determined to control them, insects engage in sexual reproduction and have a great deal of diversity in their populations. Most of the insects are killed or at least damaged by exposure to the pesticide, but there are a few that can tolerate high doses of the chemical. These pesticide-tolerant individuals then take advantage of food sources in pesticide-covered regions that other members of their population cannot access. With food in abundance, they thrive and breed more rapidly than those that were poisoned.

Eventually, under such circumstances, these surviving insects pass along their pesticide-resistant genes to the rest of the population through sexual reproduction. The pests are back, returning this time with a vengeance. Most of the population has the tolerance trait, rendering the pesticide ineffective against them.

Based on this evidence, there is no question that Darwin and Wallace's theory of evolution via natural selection is correct. Where the two scientists were mistaken was in their perception of time. They both assumed that species would always have high levels of variation in their populations and that regular encounters with environmental challenges, such as the pesticides in the example given above, would lead to some members of the varied population breeding more effectively than others. They assumed that slowly, surely and steadily the population would change. However, when Niles Eldredge and Stephen Jay Gould both carefully analysed the fossil record, what they saw suggested that the pace of evolution was anything but steady or sedate.

Eldredge and Gould were well studied in the concept of the Cambrian Explosion, and they puzzled over whether it really was an 'explosion' or just the result of the bias for hard-bodied organisms in the fossil record that we have just seen. These two palaeontologists realized that the Cambrian period was not unique in showing a sudden increase in biodiversity because similar increases have occurred since.

About 251 million years ago, during a time period known as the Permian, the fossil record reveals that a major extinction occurred, wiping out most of the species alive at the time. However, reptilian creatures survived and, suddenly, virtually overnight as far as geologists are concerned, these creatures experienced a dramatic increase in diversity. The era of the dinosaurs was born.

—

*There is no question that Darwin and Wallace's theory of evolution via natural selection is correct. Where the two scientists were mistaken was in their perception of time.*

Another extinction occurred 65 million years ago at the end of the Cretaceous period, eliminating the dinosaurs that had dominated the planet for more than 150 million years. This time, the mammals survived and subsequently diversified. Again, in a palaeontological blink of an eye, a matter of just 10 million years, mammals went from being rodent-like creatures living under the heels of the dinosaurs to large predators and herbivores thriving all around the world.

Given this undisputed evidence, Eldredge and Gould argued that evolution was more of a stop/start phenomenon, taking place in sudden bursts rather than in any sort of steady way.

These examples of extinction followed by rapid diversification look very similar to the island effect discussed in Chapter 3. We have already seen how a hypothetical pair of blue parrots was blown in a storm to an offshore island where, with plenty of food and no predators, the pair successfully bred and quickly filled the island with large flocks. But as the population grew and food became scarcer, populations of the birds began to evolve in certain locations, and particular traits made some birds more successful at surviving and breeding than others. Parrots with bigger beaks, for example, found they could eat nuts, and nut-eating parrots were soon found around the nut trees. We saw how it would not take long before the original pair of parrots would have

The discoveries made at Burgess Shale in the Canadian Rockies have been at the heart of the debate about the Cambrian Explosion.

been the ancestors of multiple species on the island. Similarly, the reptiles that inherited the Earth after the Permian extinction and the mammals that did so after the dinosaurs died out were exposed to a landscape loaded with resources and few competitors. In other words, the extinction of many animal species wiped the landscape clean, setting the stage for surviving animals to diversify rapidly.

Eldredge and Gould proposed that throughout most of the history of life on Earth, living things have not significantly evolved. They suggested that organisms typically find their niche and maintain the characteristics that allow them to exploit their niche for as long as they can. Once they have the ideal characteristics for their way of life, evolution is paused and change almost grinds to a halt. During this pause, or stasis, as they called it, there are tiny evolutionary wobbles with some variation appearing in the population at a steady rate, but most of the variation does not lead to change because the species already has a near-ideal form for its niche.

Under such circumstances, the population remains pretty much unchanged. The two researchers argued instead that when change does happen, it takes place in a burst. Following mass extinctions or similarly dramatic geological events, where large portions of the species on Earth die out, the sudden availability of untapped resources on the planet jolts evolutionary change dramatically. Gould named this phenomenon 'punctuated equilibrium' and suggested that long periods of stasis interrupted by sudden periods of major change were probably typical throughout the history of life on Earth.

Most importantly, the theory of punctuated equilibrium can be applied to the Cambrian Explosion (as well as the post-Permian and post-Cretaceous diversification events). Soft-bodied organisms living before the Cambrian period may have gone through a mass extinction that left the survivors free to exploit newly available resources. Certainly, the fractal organisms from Mistaken Point and the Australian fossil beds seem to be absent in Cambrian rocks. Whether they abruptly became extinct 542 million years ago or gradually declined is impossible to determine – palaeontologists simply do not have enough fossil beds from the period to fill out the picture. But the extinction of fractal organisms and other species from the Precambrian would have given any surviving organisms a free run of the world's oceans and food resources. With this food and space, they would have spread throughout the subaqueous landscape, just as our blue parrots filled the island.

So Eldredge and Gould left fossil collectors another puzzle to solve. What caused the mass extinction that allowed surviving life to spread and evolve so quickly? This is a particularly tricky question, not least because palaeontologists were already struggling to agree on the causes of the much more recent Permian and Cretaceous extinctions. Now they were looking for clues to explain a third extinction.

The causes of mass extinctions are some of the most hotly debated matters for palaeontologists and geologists alike. Why did large portions of life on Earth suddenly die out at specific moments in the history of our planet?

Despite the evidence available and the extensive research carried out, there is still no consensus over the two most recent extinction events. While palaeontologists agree that dinosaurs became extinct around 65 million years ago (known as the K-T extinction), there are some who argue that dinosaurs died out in a fiery explosion caused by an extraterrestrial impact. Others suggest they slowly vanished as a result of climatic and environmental change. And still others argue that massive volcanic eruptions were responsible.

To complicate matters, there is evidence to support each of the theories. In and around India, there is evidence of impressively extensive volcanic activity that perhaps pumped enough ash into the sky to darken the sun and stop plants from photosynthesizing, or filled the atmosphere with highly toxic sulphurous gases. However, near the Yucatan Peninsula in Mexico, there is evidence of a massive meteorite impact, an enormous slab of rock some 6 miles wide slamming into the Earth with unimaginable destructive power. And the theory of extinction caused by climate change is supported by evidence in the rock record indicating that temperatures and sea levels were changing.

There are a few geologists and palaeontologists who are coming to accept that the extinction of the dinosaurs was the result of several causes that took place all at once – perhaps they were even interconnected – but there is still a lot of disagreement and debate, which will most likely endure for many decades to come.

Given that the causes of more recent mass extinctions are not agreed upon, it is hardly surprising that whatever happened before the Cambrian Explosion is something of a mystery. It is even possible that the sudden burst of evolution was not caused by a mass extinction at all but, instead, by the appearance of traits in animals that allowed them to take advantage of the environment in ways that gave them an edge over all other species. This concept provides a real and possible explanation because the evolutionary explosions that follow extinctions are created by the availability of opportunity rather than by the extinctions themselves. Therefore, it is reasonable to suggest that if traits came into existence that gave animals dramatic new opportunities, these traits may have led to a diversification event like the Cambrian Explosion.

Whatever caused the Cambrian Explosion, the clues that suggest sudden diversification had long been known. The scientists needed only to contemplate the fossils themselves and ask a crucial question: why did the emergence of hard parts in animal life appear so suddenly?

———

We have already met the elongated oval organism *Kimberella*, which swam in the seas before the beginning of the Cambrian period. It is the earliest example of an animal with hard body parts, in this case an elongated plate of shell on its upper side, which may have been used to tether its muscles and improve its ability to move.

It is entirely possible that, even before the Cambrian Explosion, movement and hard parts were appearing for the sake of food gathering. The ability to move around and collect food rather than having to wait for it to float by was a major change from what had been the norm among early animal species, and a tremendous advantage.

However, the hard parts that cover the bodies of many animals today are not only used for muscle attachment. They are also essential for protection. Whether they are on clams, insects or armadillos, hard coverings make it much harder for predators to take an easy bite.

Hard parts are not always flat shells or armour plating. The external hard parts of porcupines and sea urchins are sharp, toughened spines that stick out from their bodies, making it hard for predators to get at the animal inside and, at the same time, visually announcing to the world 'not for consumption'. Imagine seeing the world from a lion's perspective as it patrols the savannah hunting for food. It hears a rustling in the grass and sinks low, creeping forward to sight its prey. It spots a group of porcupines gnawing at the bark of a tree. Now would you attack this needle-bristling rodent or try to catch one of the newborn wildebeest that you can hear bleating nearby? The sight of the porcupine's spines provides a powerful incentive for predators to search elsewhere for food.

Hard parts on an organism can also be used as weapons for attack or defence. In the modern world, the most common defensive weapons on animals are horns or antlers. Male deer grow the largest rack of antlers possible to show their strength to other males. Those with very large racks are dominant and, in turn, have an opportunity to breed with a greater number of females. This encourages larger and larger racks of antlers to appear in the population because females are choosing to breed with the males with the largest antlers and not breeding as often with those that have smaller ones. With this form of natural selection, which is called sexual selection, only the genes for the largest antlers are passed on to future generations.

But the antlers are not just for show. They can also be used to threaten other animals. Male deer often aggressively challenge one another with their antlers and sometimes actually come to blows. Indeed, it is not uncommon for male deer to die as a result of wounds inflicted during antler combat with other males.

Deer can also use their antlers to threaten would-be predators. Although their heavy hooves are far more dangerous than a large rack of antlers, no doubt they make an attack all the more difficult.

The fossil record also shows that hard parts can take on much more interesting defensive forms. The dinosaur *Stegosaurus*, for example, had a tail covered in sharp spikes. Its teeth suggest that it ate nothing but vegetation, so the vicious tail spikes would not have been used for hunting or attack.

The current theory is that *Stegosaurus* was equipped for defence; it could swipe would-be predators with its pincushion tail and cause tremendous damage. Indeed, numerous palaeontologists have had a great deal of fun trying to calculate just how much force *Stegosaurus* would have been able to bring to the end of the spikes with a solid swing of its tail at a predator. The answer appears to be a minimum of 360 newtons - for comparison, the head-on tackle of an American footballer exerts a force of around 1,921 newtons.

So it appears that the evolution of hard parts probably had the dual purpose of improving movement and defence. Intriguingly, the first burrowing animals such as the *Markuelia* may have also received an additional benefit from their newly developed digging skills. We know that animals today do not just burrow to find food; many burrow to make it more difficult for predators to hunt them down and attack them. Therefore, those first burrows could also have been perfect hiding places from lurking hunters.

It is important to realize that while burrowing and hard parts both play a modern role in defence from predators, this does not mean that these characteristics were used for protection when they first evolved. Many traits found in organisms are driven by natural selection to fill one role and then, by chance, end up proving useful in other ways.

———

What drove hard parts and burrowing to evolve? It is entirely possible that they did so initially for movement and feeding, and when the threat of predation became significant, they took on a secondary protective role. Alternatively, the reverse may have been true, with predators first driving the evolution of defensive structures that later helped the animals to feed, move, breathe or evolve further.

A return to the Burgess Shale and an examination of the fossils found there make one thing quite certain: by the time this mudstone was formed just over 500 million years ago, predation was rife. Many of the fossils from the Burgess Shale have features that can only be interpreted as defensive.

The hard exoskeletons of trilobites ensured their abundance in the fossil record and protected them against predators. This specimen was found in the Burgess Shale.

*Wiwaxia* resembled a living fortress. Overlapping plates, called sclerites, covered its body surface and sharp spikes extended away from its body.

An interesting example of this arises with an isolated, cone-shaped fossil found in the Canadian Rocky Mountains in the late 1800s, which landed on the desk of palaeontologist William Diller Matthew at the American Museum of Natural History in 1899. Having never seen anything like it before, Matthew proposed that the fossil belonged to a group of extinct shelled animals known as hyolithids. When Charles Walcott later discovered the Burgess Shale fossils, he found clusters of similar cone-shaped fossils and realized that the individual cone studied by Matthew was not an animal in its own right, but part of a large creature made up of plates and cone-shaped spines. Walcott proposed that the animal, which was named *Wiwaxia*, was a worm of sorts. But, in 1985, palaeontologist Simon Conway Morris at Cambridge University countered Walcott's suggestion. He argued that *Wiwaxia* did not share enough characteristics with worms and actually belonged to a strange group of animals no longer alive today.

—

*With such a bizarre animal, it should come as no surprise that Conway Morris's interpretation of the evidence has been challenged many times. One of the key criticisms of his theory is that it required* Hallucigenia *to use spines as walking appendages.*

Regardless of *Wiwaxia*'s relationship to other organisms, since the discovery of the Burgess Shale fossil beds, more than 138 different specimens of *Wiwaxia* have been found. There is no getting around the fact that it looks to have been a living fortress. No more than 5 cm (2 in) in length, it was covered from tip to toe in armour. Tiny plates, called sclerites, ran along the body surface and overlapped one another, like panels on a medieval suit of armour. These plates provided *Wiwaxia* with considerable protection. It also possessed spines that would deter any predators considering the animal for a meal. Running down the back of *Wiwaxia*, pointing outwards and upwards, were sharp spikes that extended to 5 cm (2 in) away from its body.

Small *Wiwaxia* specimens typically have only a few spines, but larger specimens have been found with up to 12 spines on each side. It appears that if *Wiwaxia* grew larger as it aged, which is the case with most animals, it steadily improved its defences. This creature took no short cuts in showing that any attack would simply result in a choking mouthful of crunchy shards.

However, protection does not come without a cost. For a start, this prickly plating would have been heavy for the relatively small *Wiwaxia* to carry around and it would not have been able to move fast. More importantly, everything that an organism builds on its body requires resources. The old maxim 'you are what you eat' is not far from the truth. Compounds that are consumed get transformed into materials for the body to use. The more defensive structures that are built, the more resources an animal must eat. All these plates and spikes would mean that *Wiwaxia* would have had to consume much more food than a small 5-cm (2-in) worm possessing no such armour.

*Wiwaxia* was not alone in spending vast resources on defence. In 1979, when Conway Morris was re-examining the fossils that Walcott collected in the early 1900s from the Burgess Shale, he came across a bizarre animal with sharp spines on its body. Conway Morris could not determine what it was. It had five legs and lobes on its back, which may have been used for feeding. It had a roundish structure on one side, but if this structure was a head, it lacked the normal characteristics. There was no obvious mouth, no eyes or other sensory organs like antennae. Walcott had initially classified it as a worm, but Conway Morris disagreed.

What kind of worm has no head, five legs and eats with its back? He argued that it needed to be classified as something truly different, and he called it *Hallucigenia* - a creature you could picture only if you were hallucinating.

The question of whether *Hallucigenia* had a head or not is nowhere near as complicated as which way was 'up' for the animal. *Hallucigenia* clearly had seven tentacles on one side of its body and seven pairs of spines on the other. Six of the tentacles were paired with six of the spine pairs, and one tentacle sat in front of the rest of the spines, apparently unpaired and on its own. In addition to the spines and tentacles, there were six smaller tentacles that were set behind six of the larger ones. And, as if this mess of tentacles and spines was not already enough, the animal had a long structure on its tail end, which seems to have been both flexible and tube-like.

Conway Morris theorized that the spines were stilt-like legs that allowed *Hallucigenia* to position its tentacles where they could snatch food out of the water. If the tentacles were used to pass food to some sort of a mouth on the proposed head structure, they would have had to pass food among themselves to the single tentacle at the front of the body, which was the only one that could reach the mouth. This struck Conway Morris as unlikely (simply because it would have been highly inefficient), and he suggested instead that each tentacle actually functioned as a mouth.

With such a bizarre animal, it should come as no surprise that Conway Morris's interpretation of the evidence has been challenged many times. One of the key criticisms of his theory is that it required *Hallucigenia* to use spines as walking appendages. This is not seen in any living animal and seemed to some palaeontologists as unlikely.

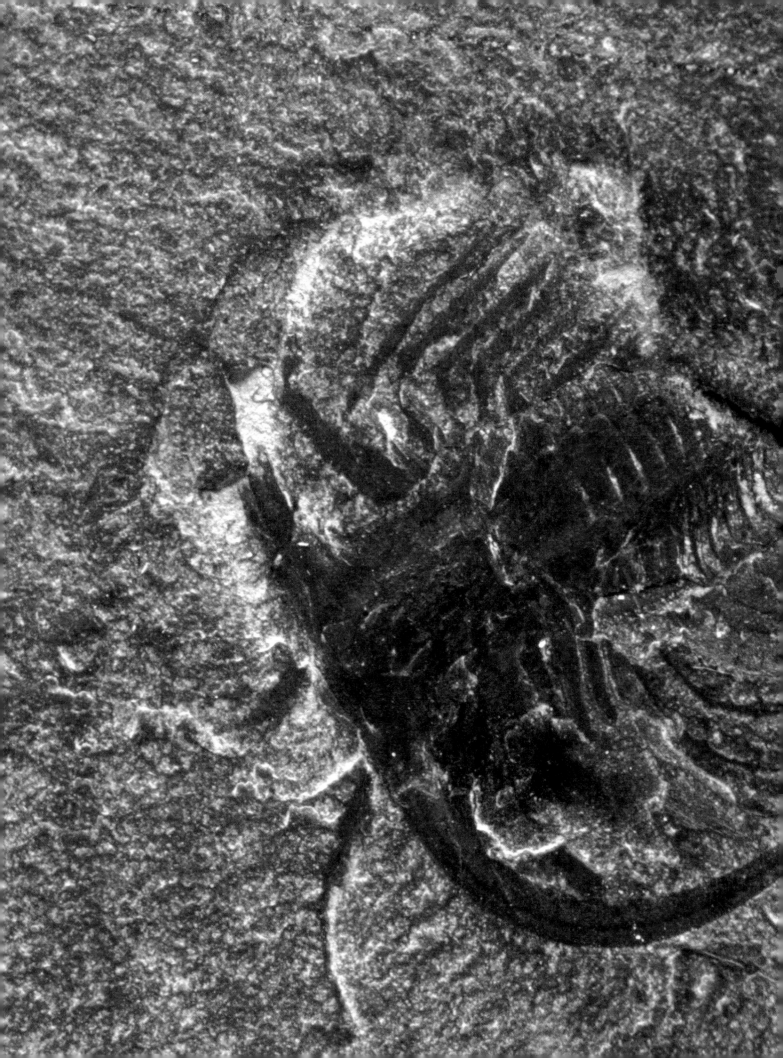

# The Marrella mystery

The Cambrian Explosion, which arose from the accelerated rate of evolutionary change in animal life, gave birth to some of the most weird and wonderful creatures ever to grace this planet. The Explosion appears to have led to some incredible evolutionary experimentation, as well as many body designs and structures that proved to be inefficient or useless and were ultimately eliminated by natural selection.

It is easy to understand how Simon Conway Morris and many palaeontologists could have mistakenly studied *Hallucigenia* upside down for so long and not noticed. During this period, many of the animals preserved in the fossil record are so unusual to us that the idea of an organism with no head and five legs was not as risible as it seems.

The trouble that scientists are having in classifying *Marrella* highlights the bizarre nature of the Cambrian fauna. The animal clearly had a head shield, which palaeontologists would expect to be hard and made from a mineralized material such as calcium carbonate. But specimens from the Burgess Shale show beyond any doubt that the shield was soft.

Moreover, *Marrella*'s body is unlike anything known to science. Its body was clearly segmented, with fossils showing about 25 segments in total. On many segments there were legs and gills, perhaps a bit like those of lobsters, and there is evidence that the segments had antennae.

Walcott proposed that *Marrella* was an unusual member of the trilobites, a group of armoured animals that became common in the late Cambrian. But others have since disagreed and suggested that it was a member of the crustacea, the animal group that crabs belong to. Other scientists suggest it was a member of the chelicerates and more closely related to spiders and scorpions. Nevertheless, the experts all agree that this animal, like so many others found in the Burgess Shale, sported defences. So, the question is, who were their predators?

*Marella splendens*, a primitive arthropod fossil from the Burgess Shale rocks. This is the most abundant of Burgess Shale fossils.

In 1991, Lars Ramsköld, a Swedish dentist-turned-palaeontologist working at the Swedish Museum of Natural History and palaeontologist Hou Xianguang at the Chinese Academy of Sciences proposed an alternative suggestion for how *Hallucigenia* lived. After discovering and analysing multiple *Hallucigenia* specimens from Cambrian fossil beds in China, they offered a simple explanation: Conway Morris had *Hallucigenia* upside down. They argued that instead of walking on its spines, *Hallucigenia* used its spines for defence and spent most of its time sticking its tentacles into the mud. They also argued that Conway Morris's suggestion that the animal had a head was incorrect – the roundish structure he was noticing was just a stain rather than a fossilized piece of *Hallucigenia*'s anatomy.

Ramsköld and Xianguang's theory for how *Hallucigenia* lived allowed palaeontologists to breathe a sigh of relief. *Hallucigenia* had been transformed into a mud feeder that stuck its tentacles into sediment in search of food. There was no longer any need to explain how or why an animal living at the bottom of the ocean would evolve stilts for legs, especially as no such animals alive today have such an arrangement. The theory also suggested that the spines were for defence rather than for walking. This made sense, since *Hallucigenia* was living in the same environment as *Wiwaxia* and was roughly the same size. If *Wiwaxia* needed serious defences, it was logical to assume that *Hallucigenia* without spines would have made a tasty morsel for the predators prowling early Cambrian seas.

Even the most common animal found in the 505-million-year-old Burgess Shale, a creature that was discovered during Walcott's first excavation in 1909 and named *Marrella*, shows the evolution of protective structures.

Like *Hallucigenia* and *Wiwaxia*, *Marrella* was a tiny animal that measured a few centimetres in length. But though small, it was not vulnerable. Its head carried a shield with two spines on each side that pointed backwards. Walcott thought the shield must have been stiff, but it was not made of hard minerals like calcium and it left an impression fossil in the soft Burgess Shale sediment instead of actually being preserved. To this day, nobody has worked out what the head shield and associated spines were made of. Nor have palaeontologists agreed on what sort of animal *Marrella* actually was, so they cannot compare its armour with the defences on other species, either living or long dead and fossilized.

The strange body parts seen in fossils found in the Burgess Shale, such as the sharp spines of *Hallucigenia*, suggest that predators were becoming more prevalent.

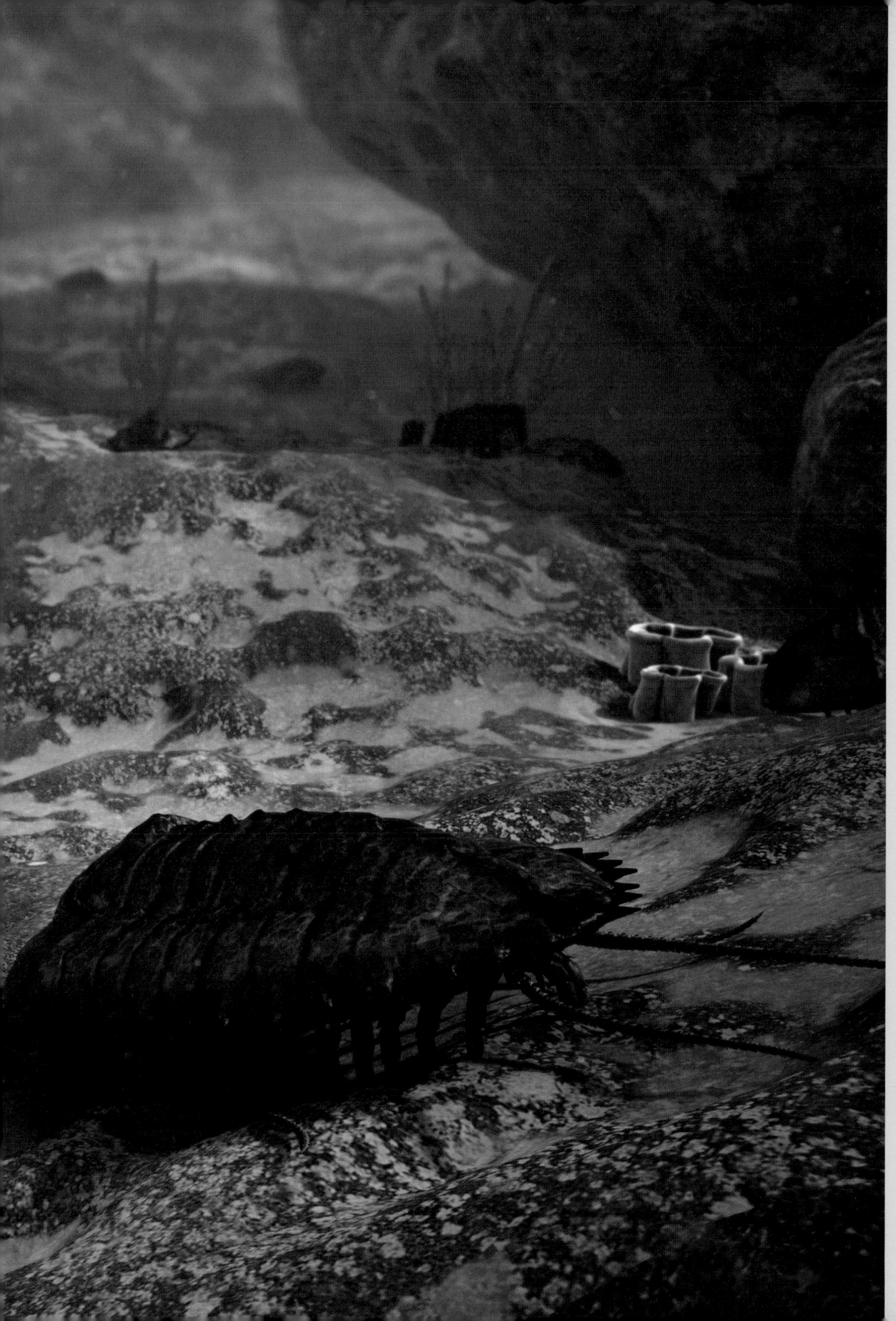

# CHAPTER 8 • HUNTER AND HUNTED

ON MOUNT ISSAMOUR in Morocco, excavating fossils has become a major industry. The rocks here, which were laid down about 150 million years after the Burgess Shale, contain thousands of trilobites. At the beginning of the Cambrian period, these creatures began to proliferate, evolving into all sorts of forms, and for the next 250 million years they were probably the most advanced forms of life on the planet. They're called trilobites because their bodies were in three sections: a head, a middle section and a tail. But there was much greater diversity in the trilobite population than the name might imply – around 17,500 known species of trilobite existed, and new fossilized species are still regularly discovered.

Moroccan trilobites are sold to collectors around the world, with the best-preserved and rarest specimens fetching thousands of pounds. Extracting the fossils from the Atlas Mountains requires perseverance, as they are scattered far apart in the hard rock encasing them. But when the specimens are found, they are extraordinary. Some have features that are so delicate that it can take days, even weeks, to prepare a specimen fully. Every particle of rock must be carefully sanded away, with tools similar to those used by a dentist, before the specimen is exposed for the first time in millions of years. It's a specialized job that requires enormous patience and skill.

The hard exoskeletons of trilobites ensured their abundance in the fossil record but they were also key to their survival. The armour of the trilobites, along with their shields, spines and spikes that were so common in the Burgess Shale and other early Cambrian fossils, make it evident that predation was rife by this time. But trying to pinpoint where and when the threat of predation arose is not so simple. Laboratory experiments point to the existence of predatory behaviour way back in the earliest days of life on Earth. Artificial proto-cells, little more than strands of genetic material encapsulated in an oily (lipid) sphere, demonstrate certain predatory characteristics as they capture smaller blobs of lipid to become a larger structure.

We have also speculated that some single-celled organisms gobbled up other cells long before multicellular life appeared. The mitochondria in modern animal cells, and the green chloroplasts in

**Preparing trilobite fossils in Morocco is a highly specialized job as each particle of rock must be sanded away.**

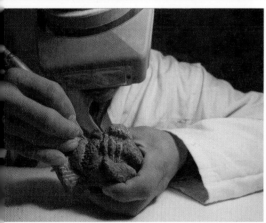

plant cells may be the vestiges of captured cells that were enslaved rather than consumed by predator cells.

However, the fossil record for single-celled life is extremely sparse, meaning that it will probably be impossible to identify single-celled predators for certain. But we can say from the fossil record that if predation did exist before the Cambrian period, it was rare. Among multicellular life, none of the fractal frond-like organisms are thought to have been predators because they had no mouths, guts or teeth. Likewise, while some sponges today are predatory, the majority are not, and researchers do not suspect that carnivorous sponges evolved in the early history of life. Therefore, it would be preposterous to suggest that *Wiwaxia* and *Hallucigenia* developed their formidable spines as protection against carnivorous sponges.

Cnidarians are more plausible candidates for being the first multicellular predators, and jellyfish are certainly dangerous predators in modern oceans. They are capable of catching and killing fish with the stinging cells on their tentacles. If their ancient relatives had evolved that predatory habit early on, they could have posed a threat to the likes of *Kimberella*, *Dickinsonia* and early worms, though they would have been powerless against an animal as well armoured as *Wiwaxia*. Unfortunately, since the soft bodies of cnidarians tend to fossilize poorly, it is hard to determine whether any of the early cnidarians in the fossil record were hunting multicellular animals or not – we will have to wait for the discovery of a cnidarian fossil with some freshly caught prey in its tentacles.

So we will probably never know definitively which specific group of organisms first started to hunt or when exactly predation began, but the sudden appearance of diverse hard parts at the start of the Cambrian period and the clear presence of heavily defended organisms in the Burgess Shale suggest that between 543 and 505 million years ago multicellular predators evolved. Most importantly, fossils excavated in the Burgess Shale have revealed – and almost all palaeontologists agree on this point – the first animal to hunt and kill its food.

———

The discovery of this first predator was made by Joseph Whiteaves, a British palaeontologist who was living in Canada. While studying Canadian rocks in 1892, Dr Whiteaves came across a curious fossilized animal that looked very much like a shrimp or the slightly curved tail of a lobster. He did not know what to make of the finding because he could not see any evidence of a gut inside it, so he named it *Anomalocaris*, which effectively meant anomalous (or odd) shrimp. He had no idea that it was a predator and never suggested that possibility in his writings.

Almost 20 years later, unaware of Whiteaves's discovery, the American palaeontologist Charles Doolittle Walcott began his extensive work in the Burgess Shale fossils. During his excavations,

# The Mystery of Hallucigenia

" I'm extremely fond of little *Hallucigenia*, not just for its unusual appearance but also for its enduring ability to provoke scientific dispute, ever since Charles Walcott discovered the first specimen in 1911.

Walcott originally described it as a polychaete worm, from a group of bristly worms which include lugworms. He named it *Canadia sparsa*. When the palaeontologist Simon Conway Morris carefully reviewed Burgess Shale specimens in 1979, he decided that Walcott's find was not a polychaete at all, but a creature so unique it deserved its own genus, *Hallucigenia*, named for its 'bizarre and dream-like appearance'.

*Hallucigenia* became a key icon for the Burgess Shale and for all the curious animals that emerged during the Cambrian Explosion. Much like the five-eyed *Opabinia*, *Hallucigenia* was thought to be an evolutionary experiment with no descendants.

It later emerged that Conway Morris's interpretation was mistaken. He drew *Hallucigenia* upside down, strutting stiffly about the Cambrian seabed on seven sets of spiked limbs, with a single row of tentacles wafting around on its back, far too short to reach its bulbous head.

Modern-day analysis of the fossils has shown that *Hallucigenia*'s tentacles were in fact paired legs and its spines were situated on its back, in order to protect it from predators. We now realise that Conway Morris had drawn *Hallucigenia* not only upside down, but back to front! The bulbous head was in fact the rear end of the creature. With hindsight, it's easy to laugh at this sort of mistake, but misinterpretations of fossils occur rather frequently, especially if specimens are relatively few in number or haven't preserved well.

Palaeontologists also recognised that *Hallucigenia* should not have been placed entirely on its own as an evolutionary mishap, but that it was probably a very distant relative of the modern-day *Peripatus* velvet worms.

In 2002, yet another theory about *Hallucigenia* was put forward. A Canadian palaeontologist called Desmond Collins noticed that all of the 30 or so *Hallucigenia* fossils seemed to fit into one of two categories, heavy or slim-line. He speculated that these might represent males and females of the species.

You would think that almost a century on from Walcott's original discovery, scientists would have reached firm conclusions on exactly what *Hallucigenia* was, but this enigmatic little creature continues to perplex researchers and stimulate radical theories. It's an interesting example of the role that debate, and reinterpretation of evidence, plays in science."

Walcott found a circlet of plates fossilized as impressions in the fine sediment. Due to their circular appearance and somewhat radial symmetry, he assumed that they belonged to a cnidarian of some sort and named the creature *Peytoia nathorsti*.

When Simon Conway Morris further explored the Burgess Shale, he found a sponge-like animal associated with the shrimp-like *Anomalocaris*. He assumed that the two animals had died and fallen into fossilization together, and he named the sponge *Laggania*.

Rather remarkably, all three palaeontologists were mistaken; they believed they were looking at unique animals. It wasn't until 1979 that Yale University palaeontologist Derek Briggs suggested that *Anomalocaris* was not an individual animal at all. Instead, he argued that it was the appendage of a much larger organism related to insects and crustaceans. He also hinted that he expected a large animal with appendages like *Anomalocaris* to be found one day in the Burgess Shale. He was soon proved right.

Two years later, in 1981, Harry Whittington, a palaeontologist at Cambridge University, was working with Briggs on an unidentified fossil from the Burgess Shale. As they prepared the specimen by gently clearing off bits of sediment and exposing the imprint more fully, they recognized the telltale circlet of *Peytoia nathorsti*. Yet this *Peytoia nathorsti* was special. It was not on its own; it had *Anomalocaris* specimens associated with it. This was most unusual and led the team to examine their strange find more closely.

They found that the *Anomalocaris* individuals were not just next to the specimen of *Peytoia nathorsti*, they were literally attached to it. This suggested that Briggs's theory was correct. *Anomalocaris* was not a strange, shrimp-like creature with no digestive system, but rather a type of arm sticking out from the body of a larger animal. *Peytoia nathorsti* was not an independent creature either – Briggs and Whittington realized that they weren't staring at a cnidarian but into a rounded mouth. Closer examination of *Peytoia nathorsti* fossils since then has shown that they were often attached to, or frequently associated with, fossils of the *Laggania* sponge that Conway Morris had discovered.

A realization set in: *Laggania* was also not an individual at all but a body part of something larger. Whittington and Briggs's prepared fossil clearly showed that *Anomalocaris*, *Peytoia nathorsti* and *Laggania* had all been misidentified; they were, in fact, all body parts of a single, large animal. Based upon the scientific rules for the naming of organisms, the new single species took the name that its first discovered body part was given: *Anomalocaris*.

The size of *Anomalocaris* was one of the reasons that palaeontologists failed to spot it as a unique organism because nothing else grew as large as it in Cambrian waters. *Hallucigenia* and *Wiwaxia* were both

As far as we know,
*Anomalocaris* was the
first big predator on Earth.

just a few centimetres long. *Anomalocaris* was up to 70 cm (2 ft) in length, making it a behemoth of its time.

Extensive study of its body has also shown that, unlike *Hallucigenia* and *Wiwaxia*, for which detecting the presence of heads, eyes and mouths has often been difficult, *Anomalocaris* definitely had two eyes at the front of a very obvious head and a mouth of many overlapping plates. On these plates were numerous tiny spikes oriented towards the centre of the mouth. The structure of the plates and spikes suggests that *Anomalocaris* probably opened and closed its mouth by sliding the plates over one another. It would have been able to exert pressure on any food objects in the way by using its spikes. Moreover, in front of the mouth it had two long, arm-like structures – the 'odd-shrimp' appendages first discovered by Whiteaves – covered in tiny barbs.

You only have to look at the barbed, grasping arms and the mouth with pointed teeth to see that this was a killer. The animal's size is also a clue to its hunting behaviour. Most predators tend to be bigger than the animals they hunt: grizzly bears do not hunt adult elk; they go after youngsters. Similarly, sharks eat fish, snow leopards eat hares, and falcons catch sparrows. Even without all the weaponry, the size of *Anomalocaris* immediately points the finger at its predatory nature.

*You have only to look at the barbed, grasping arms and the mouth with pointed teeth to see that this was a killer. The animal's size is also a clue to its hunting behaviour. Most predators tend to be bigger than the animals they hunt: grizzly bears do not hunt adult elk; they go after youngsters.*

Many palaeontologists speculate that *Anomalocaris* must have had the ability to snatch food items with its arms and put these items into its toothy and rounded mouth of sliding plates. Its shape also suggests that it was most likely a swimmer: it had a tail that was probably flexible enough to generate some propulsion, as well as a series of fin-like structures along the edges of its body that it could probably flap to propel itself around.

In other words, *Anomalocaris* is the first-known predator on the planet, a giant of its time and, undoubtedly, the scourge of the Cambrian world.

But what sort of food was *Anomalocaris* eating? Certainly not algae or filter-feeding, single-celled plankton from the water. Yet evidence that it was hunting animals like *Hallucigenia* and *Wiwaxia* is scant. If this were the case, it would seem likely that broken, bitten or partially digested fossils of these animals would at least occasionally turn up in association with *Anomalocaris*. Instead, it looks like it may have been hunting another group of organisms that became extremely common during the Cambrian period: the trilobites.

———

Trilobites were heavily armoured animals that wore their skeletons on the outsides of their bodies, as insects, crabs and lobsters do today. As we have already seen, it is an open question whether their external skeleton was initially being used for support and muscle attachment or for protection, but there is little doubt that by the time that *Anomalocaris* was common in the Cambrian oceans, their armour, or exoskeleton, was playing a key role in keeping them alive.

Professor Richard Fortey, a palaeontologist at the Natural History Museum in London, believes that the exoskeleton was the key to the success of the trilobites. 'The trilobites did almost everything you possibly can do with an exoskeleton,' Fortey says. 'I think that their skeleton is what gave them an advantage. They were protected so they could do all kinds of interesting things. They could evolve to grow spines, or to be flat like pancakes. They could protect themselves by developing a thick exoskeleton with pebbles all over it. It was a great advantage to them just as it is to crabs and lobsters living today, which of course came much later into the evolutionary journey than the trilobites. So they utilized the virtues of having a tough exoskeleton to radiate into all kinds of ecological niches.'

The smallest trilobites were just one millimetre long. But there is huge variation among the thousands of specimens found. This one, uncovered in Morocco, is one of the larger specimens.

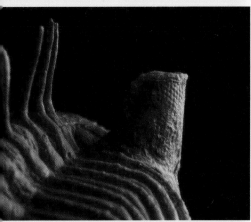

# Trilobites

" They might seem like a strange choice of animal to be passionate about, but I've always had a soft spot for trilobites. Very few people really appreciate how complex, varied and beautiful these creatures were, or how long they roamed the Earth. Trilobites persisted in the fossil record for an astonishing 300 million years – longer than the dinosaurs, and longer than mammals have lived on Earth.

Trilobites are magnificent not only for their staying power, but also for evolving to fill a whole range of evolutionary roles. Some ploughed through the sea mud, feeding on sediment and peering out above the mud with eyes on long stalks. Others cruised through the upper level of the water, scanning the seabed for prey, a little like a hawk does in the air.

Their tough exoskeletons not only protected them from predators like *Anomalocaris*, but diversified to a great extent. The variation you can see in fossils like those from Morocco is simply astounding. Some evolved elaborate curled horns; others had forked appendages they waved in front of them.

I find their multi-faceted eyes especially mesmerising. Trilobites' eyes were similar to the compound eyes of insects today. They were made of thousands of tiny lenses that each collected a small snapshot of the world around them. Presumably their brains processed these many small images to work out what was happening in the world around them just as insect brains do.

Trilobite eyes differed from insect eyes in one important way. Insect eyes are made of fragile materials and decay into nothingness when the insect dies, but trilobite eyes were made of a hard material, calcite. This is the same mineral that is in sedimentary rocks like limestone and marble. It preserves perfectly, meaning that when looking at a trilobite eye under a hand lens you can make out in wonderful detail hundreds upon hundreds of neatly packed, hexagonal compartments. These compartments make up an eye that saw the world 500 million years ago. It's a wonderful thought.

I suppose for me the fascination with trilobites arose as a boy. Though I delighted in every new fossil find, I got pretty blasé about the ammonites and belemnites, which were quite common in the Jurassic rocks of Leicester. I longed to crack open a rock and see something more exotic like a vertebrate, but would have happily settled for a trilobite. The best place in Britain to find trilobites is Wales, but that was a long way away from Leicester and somewhere we only visited on holiday. I suppose if I'd grown up in Wales I might have thought that trilobites were terribly dull and prayed for an ammonite!"

Unlike *Anomalocaris*, trilobites have been known about for quite a long time. They were first named by Johann Ernst Immanuel Walch, a German theologian and philosopher, who wrote about the fossils he discovered in 1771. Trilobites are common and easy for even amateur fossil collectors to spot because their hardened armour fossilized readily in ancient oceans and did not require the fine sediments that make the Burgess Shale such a special fossil-hunting ground. Indeed, their armour fossilized so often and so well that, were it not for the impression fossils of soft-bodied creatures found at the Burgess Shale, it would be easy to make the mistake of assuming that trilobites dominated the planet on their own for millions of years.

Thanks to their substantial presence in the fossil record, trilobites offer something valuable that the Burgess Shale's soft-bodied animals cannot: a chance to see obvious injuries. Such an abundance of trilobite fossils have been found, many of which are the same species, that it is possible to compare individuals and see trends among them. In 1999, Christopher Nedin, a palaeontologist at the University of Adelaide, noted one characteristic that he found particularly fascinating in a trilobite unearthed from early Cambrian fossil beds in South Australia. The trilobite had damage to its armour. Based upon careful analysis of the injuries and comparisons to the mouth and appendages found on *Anomalocaris*, Nedin proposed that this was strong evidence indicating that *Anomalocaris* was attacking trilobites.

Imagine the scenes in the oceans millions of years ago: a trilobite scurries across the bottom of the ocean like an oversized, armour-plated woodlouse. It is moving fast, seeking cover, because it has spotted *Anomalocaris* powering out of the murky shadows. But it's too slow and the arms of the fierce predator shoot forward and snatch the trilobite off the sea floor. But what happens next if *Anomalocaris* couldn't crack through the trilobite's solid protection?

Based upon the injury he was studying, Nedin pictured *Anomalocaris* using its two front appendages to wrap around the trilobite body and flex the hapless animal back and forth. This flexing of the trilobite, he suggested, took advantage of a weakness in early trilobite body structure and created an opening in the tough body armour. Once opened, *Anomalocaris* would have been able to get at the soft and edible interior of its prey.

———

However tempting Nedin's argument might be, it is not the only proposal for how *Anomalocaris* hunted. Justin Marshall, a marine biologist at the University of Queensland, suggests that *Anomalocaris* might have attacked in a rather more spectacular way. With a diving mask and air tanks, he drops beneath the water to pull out a specimen in order to demonstrate a hunting technique that may date back more than 500 million years.

# Looking at Anomalocaris

"When I began hunting for fossils as a boy, palaeontology was fairly limited to the act of collecting fossils, finding out what they were and putting them on a shelf to be admired. The problem with this way of doing things is that by just examining a fossil with your eyes you can miss so many things.

Palaeontology has undergone something of a technological revolution in recent years. Now, in specialized laboratories, you can examine fossils with hugely powerful X-rays or with microscopes that use electrons instead of light. This not only allows you to look at a fossil in extraordinarily fine detail but to actually explore the interior of the animal and work out how it fitted together and how its organs or muscles would have worked.

An advance that has been really thrilling for me has been the incorporation of computer graphics into this analytical process. Now you can take all the scientific information about an animal, including the exact measurements of a fossil, and put it into a computer in order to make the animal climb out of the rock, come to life, and actually move. In this way, it's possible for me, a person who has come back to these fossils many times over the last 50 years, to see them in an entirely new light.

No matter how imaginative you are, there is no better way to recreate an animal than with computer imaging. Not only can you see the animal, you can apply the physical rules of the real world. You can factor in the weight of water or the air pressure, add in Newton's laws of gravity and force, and in this way you can work out, with scientific justification, exactly what this animal would have been capable of, and what was impossible.

As well as physics, inspiration comes from the natural world. In *First Life*, we recreated what is perhaps one of the world's earliest predators, *Anomalocaris*. We can obtain clues to what *Anomalocaris* was like from an arthropod that is alive today. On the Great Barrier Reef, in Australia, there's an animal that, although much smaller, is remarkably similar to *Anomalocaris*: the mantis shrimp.

The mantis shrimp has an ancient ancestry. Fossils of almost identical creatures have been found that date back 400 million years. This animal is almost as ancient as *Anomalocaris* itself. It lurks in burrows waiting for its victims to swim within range, and then it strikes.

Looking at the fossils of *Anomalocaris* and comparing them to mantis shrimps of today, one could suggest that these are both very similar apex predators, as both have large raptorial appendages with which to grab their prey.

While recreating *Anomalocaris*, CGI experts studied the lightning-fast movements of the mantis shrimp, and combined them with the wave-like body movements of modern-day hunting fish and sharks. With these for inspiration, the team brought to life the ultimate Cambrian predator in a way that has never been seen before."

The mantis shrimp, although much smaller, is remarkably similar to *Anomalocaris*.

# To Snap or Suck

The verdict on *Anomalocaris* is not certain. Even a decade after Nedin's work with the animal and all of Marshall's comparisons with mantis shrimps, there is still considerable debate about how the great Cambrian predator behaved and what it ate.

James Hagadorn, a palaeontologist at Amherst College in Massachusetts, is one researcher who thinks the role that *Anomalocaris* played in hunting trilobites has most likely been overstated. In 2009, at a research conference in Canada, Hagadorn argued that *Anomalocaris* was probably not the voracious trilobite hunter that is depicted in almost every description of the animal – including this one here so far. Using computer models that explored and demonstrated how the mouthparts in *Anomalocaris* might have worked, Hagadorn showed that its mouth might not have been as adept at crushing, snapping or scratching.

According to the computer model, *Anomalocaris* could not completely close its mouth, and this distinctive feature would have made it difficult for the animal to chew. More importantly, it would have been very hard for the predator to bite with any sort of significant force. Indeed, Hagadorn pointed out at the conference that trilobite injuries do not look like they could possibly have been caused by a mouth like that of *Anomalocaris*.

Computer models are not alone in tempering the image of *Anomalocaris* as the aggressive trilobite hunter of the Cambrian seas. Modern predators that attack animals with hard body parts often carry evidence in their mouths of their predatory behaviour. Lions, wolves, bears, sharks and many other predators break teeth all the time when they attack prey animals that have either armour or hardened skeletons. These broken teeth sometimes show up in the fossil record embedded in the bones of prey animals, but far more often these breaks can be seen in the mouths of the fossilized predators. Indeed, it is exceedingly rare to find fossilized predators that do not have broken or damaged teeth.

Hagadorn reasoned that if *Anomalocaris* were attacking animals as heavily armoured as trilobites on a regular basis, then the predator would show some signs of wear and tear on its mouth plates or on the sharp barbs inside the plates. Yet, rather remarkably, damage from biting hard armour is never seen. The tips of the barbs inside the predator's mouth do not even appear to be worn down at all.

Given the presence of trilobite pieces in pellets, and the trilobites from the fossil record with healed injuries from biting attacks, one would expect to see quite a lot of damage on the mouth parts of *Anomalocaris*. In Hagadorn's opinion, this lack of wear and tear stood as evidence that interpretations of the animal as a predator of armoured trilobites were wrong.

So Hagadorn is suggesting a reformed character: an *Anomalocaris* who now cannot snap trilobite armour or even bite properly. Hagadorn

The teeth of *Anomalocaris* can be seen in its fossilized mouth, although how exactly they were used has been much debated.

has also observed that the inside of the mouth plates of *Anomalocaris* were often wrinkled and deformed. This suggested to him that these mouth parts might not have been very hard when the animals were alive. The mouth plates also showed fracture patterns that are similar to those that researchers see today when the semi-flexible armour on the outside of lobster and crab bodies becomes old and dried. If the mouth plates were made of the same material as lobster armour, they would certainly not have been effective at breaking trilobites open. Indeed, models using shrimp and lobster armour show that mouth plates made of these materials would have been able to withstand a force of no more than 6.2 and 13 newtons, respectively. Most Cambrian trilobites with healed injuries were strong enough to handle 3.7–37.1 newtons of force, so only the very weakest armoured trilobites would have been vulnerable to an *Anomalocaris* bite.

Drawing from all this data, Hagadorn argued that perhaps *Anomalocaris* was a predator, but a very different sort from what was being suggested. Rather than cracking open trilobites with its appendages or fiercely biting with jaws, Hagadorn suggested that it might have specialized at sucking up soft-bodied prey.

Supporting this argument, computer models show that the mouth would have most likely worked in concert with muscles that are commonly found in modern round-mouthed organisms to suck up food. The mouth plates and muscles probably functioned like a sphincter and generated suction like a vacuum cleaner.

Hagadorn argues that while *Anomalocaris* may have bitten some soft trilobites, it is unlikely to have pursued such heavily armoured prey. Instead, the predator was far more likely to have been using its appendages to comb through sediment in search of soft-bodied animals that were burrowing to avoid being noticed and eaten. Once it found these prey animals, it probably sucked them up; if it closed its mouth a bit, the small barbs on its plates probably helped to keep wriggling and slippery prey from escaping, rather than actually biting them.

So perhaps *Anomalocaris* is not the only predator that rampaged through the Cambrian period devouring trilobites. In that case, which creature made the obvious bite marks found on trilobites or spat out the pellets of pulverized trilobite armour? What other yet-to-be-discovered monsters were prowling the deep? Once again, there is no definitive answer to such questions and there have been no theories put forward that have been beyond reproach, because scientific study is an ongoing debate of the evidence gathered.

**Like the mantis shrimp that lives today, *Anomalocaris* had big raptorial appendages that allowed it to grasp prey.**

Soon he resurfaces, gingerly holding up a wire loop in which he has trapped a large pale shrimp, roughly as long as his hand. It is streaked with ruddy brown patterning and has reddish legs and tail – a mantis shrimp that is quite common on the Great Barrier Reef.

He carefully places it in a tank where its resemblance to the praying mantis that lives on land is evident. The shrimp's front legs are coiled up and it appears somewhat penitent. But this appearance of peaceful religious reflection is quickly lost because mantis shrimps are aggressive predators.

When they attack, they quickly unfurl their appendages and smash whatever animals are before them. Some species of mantis shrimp have spines on the ends of their appendages that can impale prey, while others simply bludgeon their prey by whacking them with their elbows. 'Fisherman call them thumb-splitters, because as they handle them they often get their thumbs split open,' Marshall explains.

Most remarkably, mantis shrimps are the fastest-known predators alive on the planet. The speed of their strike with those adept arms can reach 45 miles per hour, delivering the same force as a bullet shot from a gun. The great speed with which they unfurl their appendages creates heat in the surrounding water that sometimes actually boils it. 'They're really very nasty animals,' adds Marshall.

—

*Most remarkably, mantis shrimps are the fastest-known predators alive on the planet. The speed of their strike with those adept arms can reach 45 miles per hour, delivering the same force as a bullet shot from a gun.*

An *Anomalocaris* fossil (right), highlighting from top to bottom the entire fossil, mantle, tail, appendages and mouth.

From Marshall's perspective, the appendages on the mantis shrimp carry a striking resemblance to the appendages on *Anomalocaris*, and he suggests that it is likely that *Anomalocaris* hunted animals in the Cambrian oceans with similar lightning speed.

Regardless of how they might have attacked, there is other evidence that *Anomalocaris* had a taste for trilobites. Discarded pellets associated with Cambrian fossil beds have long been known to contain pieces of trilobite in them. Early palaeontologists wondered whether these pellets were like those produced in some modern animals such as owls. Owls eat their prey in large bites, and can even take down a small mouse whole. They have no teeth and depend upon a combination of acid and grinding mechanisms in their digestive systems to break up their prey. Even so, they cannot digest all of the fur and hard bits that they ingest and, consequently, they cough up pellets of undigested material, which they then spit on the ground.

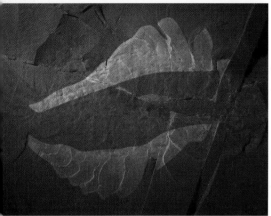
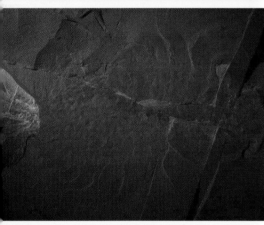

Until the realization that *Anomalocaris* was a single, large animal, nobody could figure out which creature in the Cambrian period was big enough to produce the pellets. But we now know the likely culprit: *Anomalocaris*. Nedin suggested that it was probably absorbing as many nutrients from the soft tissues of the trilobites as it could and then dumping out the inedible hard parts in a single pellet, in the same way as owls do.

———

While *Anomalocaris* is undisputedly the largest-known predator of the Cambrian period, it was not the only predator. While *Anomalocaris* cruised the waters, another predator lurked in the sandy muck of the seabed. Unlike *Anomalocaris*, this predator, which was first discovered by Charles Walcott and is extremely common in the Burgess Shale, did not move around in search of its prey. Instead, it seems to have sat still and waited in hiding.

The animal, which has come to be known as *Ottoia*, is typically little more than 70 mm (2.75 in) in length and it is often found with its body in a curved shape. From the fossils discovered, the animal seems to have spent its life underground in burrows, with just a small part of its head near the surface of the sediment. What it was doing with its head seems obvious to palaeontologists for two reasons. First, fossils of *Ottoia*'s head clearly show that its mouth was covered in sharp barbs, which would have been effective at impaling animals slithering past its burrow. Second, there are worms in the ocean today that closely resemble *Ottoia* and also spend most of their lives in hidden burrows waiting for approaching prey.

The modern worms that look like *Ottoia* and burrow in seemingly similar ways are the penis worms, which we met in Chapter 6. While their name, which describes their shape, may not conjure up an image of fear to humans, these creatures are voracious predators that sit beneath sediment and wait to stab their prey. When a succulent animal strays too close, the penis worm slams the sharp spines on the ends of their mouths upwards. They are the ultimate booby-trap predators.

Based upon the burrow and spine similarities that *Ottoia* has with penis worms, palaeontologists have long theorized that it must have behaved in the same way. Researchers also have some impressive specimens from the Burgess Shale to back up their theory. Over 1,500 *Ottoia* specimens have been found in the Burgess Shale since Walcott began his excavations; in some of the finest specimens, palaeontologists have been able to detect impressions of the worm's digestive system – and even the food inside.

It is quite astonishing – and also a validation of the tremendous level of preservation found in the Burgess Shale – that a worm's last meal can be examined after half a billion years. Yet these special fossils provide unquestionable proof that *Ottoia* was a predator. Inside their guts, palaeontologists have found numerous specimens of a shelled

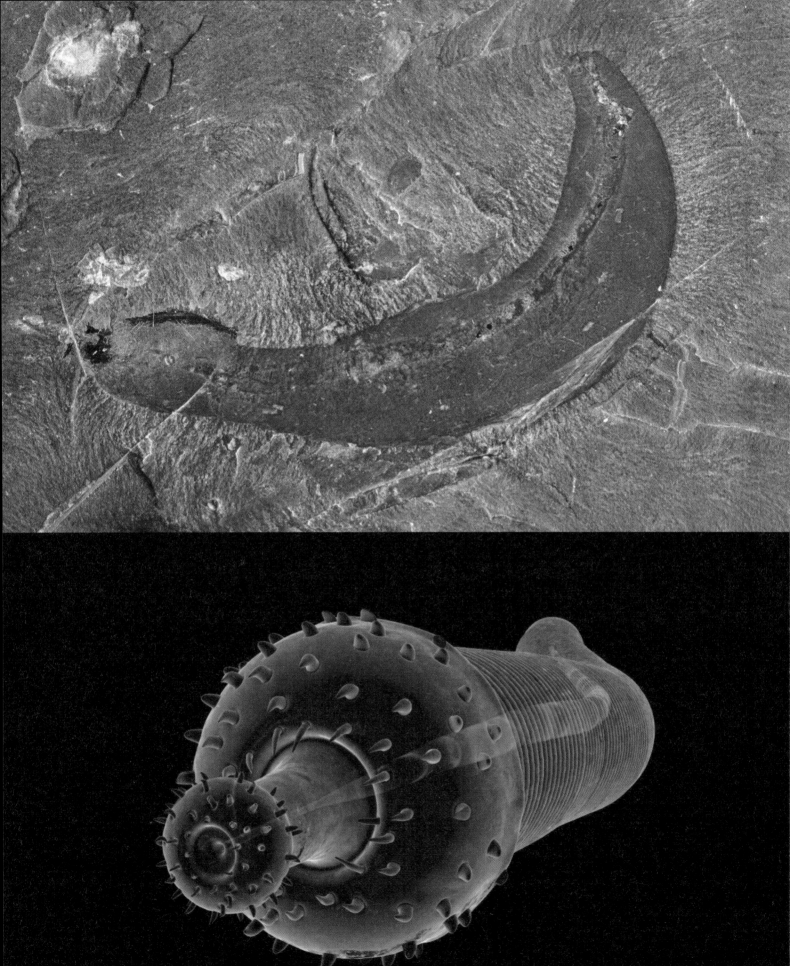

animal known as *Haplophrentis*, which was only a few millimetres in length. Intriguingly, analysis of the gut contents of *Ottoia* also suggests that they may have engaged in cannibalism, as the body of one *Ottoia* has been found inside the gut of another. Whether this activity was common is unknown, but it looks like it occasionally happened.

*Ottoia* is an extremely common member of the Burgess Shale fossil beds and, at first glance, this hints that ambush tactics were a major form of predation at the time. There is no reason to argue against such a theory, but palaeontologists are cautious in taking interpretations of *Ottoia* specimens too far.

The reason for the restraint is because *Ottoia*, as a fossilized animal, is almost too good to be true. *Ottoia* is almost perfect preservation material because it was living in soft sediment for most of its life. By dwelling in such an environment, *Ottoia* could be fossilized without the special circumstances of ash or fine sediment falling on them when they died to create impression fossils. Simply dying in their burrows would have set the stage for a fossil being formed, and the regular supply of mud at the Burgess Shale site increased this likelihood.

In other words, while *Ottoia* appears to have been extremely common, this seeming frequency might be an illusion generated by the fossil record, which has a tendency to fossilize animals living in certain environments much more readily than those living in others.

To be an apex predator, you have to be the biggest, fastest and best hunter. The upshot of that for the prey is if you are being hunted by an apex predator, you must become faster, better camouflaged and more adept at protecting yourself. You have to have defensive armour, eyes to see an attack coming and the ability to move in order to escape. The fact that *Anomalocaris* and other predators existed together would instigate an arms race among predators.

The obvious appearance of predators and the overwhelming number of animals with substantial defences, like spines and armour, seem to be tightly connected to the explosion of animal diversity at the start of the Cambrian period. But our exploration of these fascinating animals still leaves us with a burning question: why? Palaeontologists have grappled with this matter for decades and, as irony would have it, the beginnings of an answer came from outside the fossil-hunting world.

A fossil *Ottoia* and a diagrammatic representation showing the toothed head and the gut running through the animal.

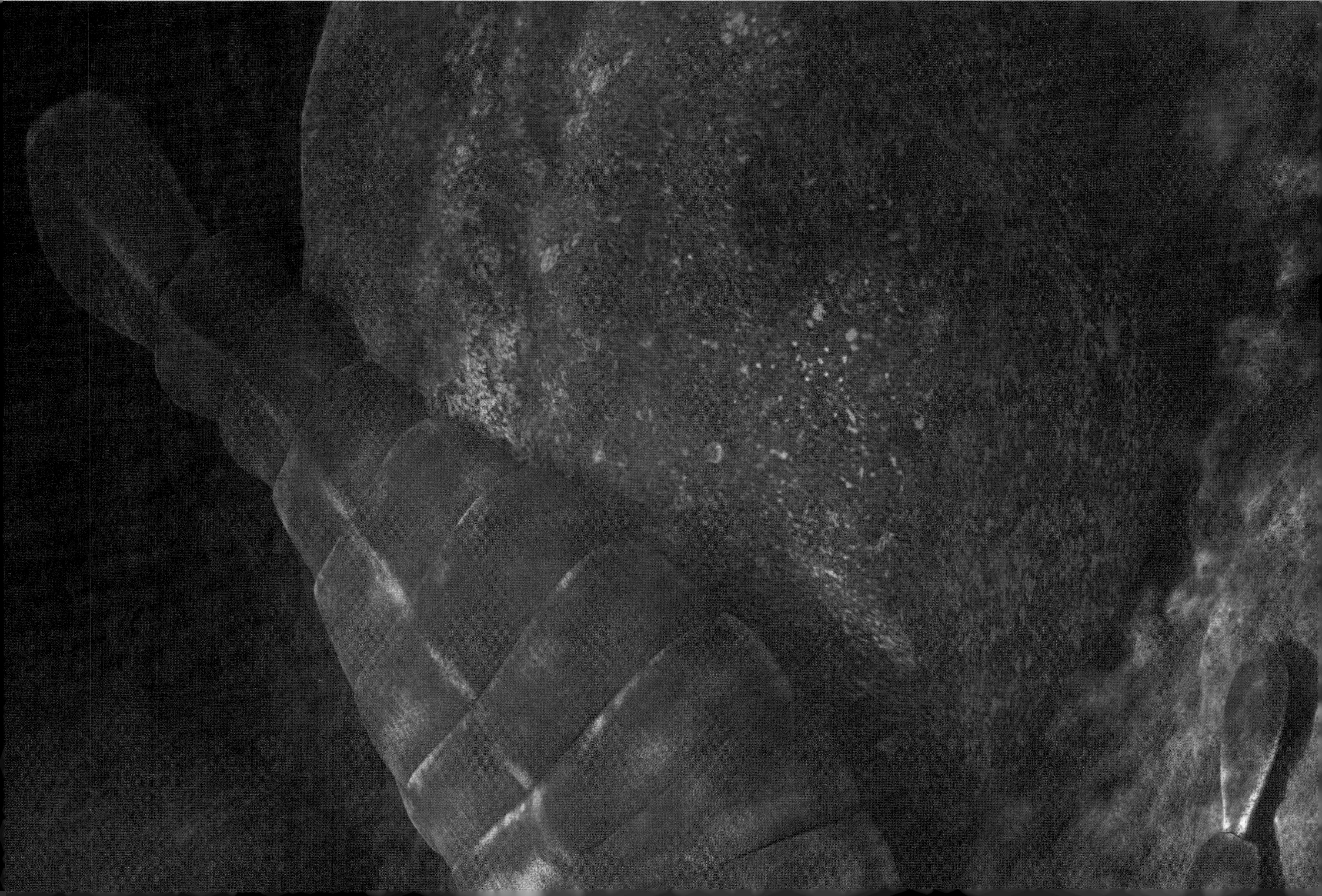

IT WAS AN insignificant barnacle that finally unlocked the mystery of the Cambrian Explosion. However, this was not the first time that the barnacle caught the attention of naturalists.

In the 1960s, Robert Paine, an ecologist at the University of Washington, investigated the roles played by different animals in a community. He wondered what result would follow after removing a single species from a local ecosystem. Like all ecologists studying animal interactions, Dr Paine knew that not every threat is made by a predator. Competition among members of the same species also plays a key role in the survival of individuals.

Take acorn barnacles as an example. These creatures barely move throughout their lives. They live on coastal rocks and wait for the tide. When water is present, they open up their hatches and collect food particles churned up by the waves. When the tide recedes, they close their doors and sit tight as the sun's heat threatens them with desiccation, and predators such as birds and carnivorous snails comb the rocks for an easy meal.

Once barnacles reach adulthood, they stick long penises out from their protective hatches and inseminate other barnacles on the rock around them. Infant barnacles, which are known as larvae, are created from this sexual activity. Unlike their adult parents, these larvae can swim, but only for a time. They must quickly find a place to settle or they will expend their energy reserves and die.

Although the danger posed by potential predators is considerable for acorn barnacles, their everyday environment holds a far greater threat. It is a matter of life and death that young barnacles settle in the best possible location, where they have access to the tide and are not too exposed to the sun, so they usually end up all congregating next to one another. However, such adjacent settling is effectively a declaration of war. From the moment one larva lands next to the other, they must compete for food so that they outgrow their neighbour and take control of the full space.

Dr Paine knew from earlier studies that the presence of predators in a locality tended to reduce the level of competition between barnacles.

These observations made perfect sense. Lots of snails and birds constantly feeding on barnacles would create more space for larval barnacles to settle without having to compete with their neighbours for survival. Yet, to Paine, there seemed to be another critical element to this. By eating barnacles, predators were opening up the environment for other rock-dwelling animals to survive as well.

To explore this idea, Paine conducted an experiment where he and his colleagues removed all the starfish from a rocky shoreline in the state of Washington. Starfish may look docile but they are actually vicious predators that attack shelled animals like mussels. These starfish were sharing the environment with crabs, kelp, mussels, barnacles and many other organisms. Before the experiment, Paine and his colleagues recorded the organisms present and calculated the relative numbers of each. After they had removed the starfish, the team regularly monitored the populations of each remaining organism. They made a startling discovery: with the loss of the predatory starfish, diversity in the ecosystem sharply declined.

Without starfish to hunt them, mussel populations boomed. Within just a few years, mussels dominated the rocks, driving out other rock-dwelling organisms. Remarkably, the presence of a predator seemed to be linked to *increased* diversity in an ecosystem. Paine concluded that as prey animals become dominant in an environment, predators respond to this by adapting to enable them to eat prey more effectively. Predators, it seemed to Paine, were critical to keeping any one species of prey animal from dominating the landscape, thereby maintaining diversity in ecosystems.

Based on these findings, Steven Stanley, a palaeontologist at Case Western Reserve University in Ohio, suggested that, because predators support diversity in ecosystems today, the same process happened in the past. Therefore, a rise in animal predators may have triggered the incredible diversification that ultimately triggered the Cambrian Explosion.

Dr Stanley theorized that if the mechanism that Paine was identifying had been present throughout evolutionary history, then when one prey species became particularly abundant in the past, natural selection would drive predators in the area to develop traits that would make it easier for them to capture and feed on that species. In principle, these newly developed traits would lead to speciation in the predators: some predators would stick with whatever animals they had always fed on and some predators would develop traits that helped them to exploit the new food source. Over time, significant differences would most likely appear in the predators, and the population would ultimately split into two predator species, each hunting their own prey. To Stanley, it looked very much like the Cambrian Explosion was the result of the rise of the first animal predators.

*Pikaia*: with the beginning of a backbone, this creature is the ancestor of all veterbrates, including us.

Geerat Vermeij, a palaeontologist at the University of California, Davis, provided an explanation for how predation might have fuelled a diversification event as large and as fast as the 10-to-20-million-year explosion at the start of the Cambrian period.

Dr Vermeij had a long history of studying shelled animals and was fascinated by the fact that many snails, clams and limpets carried scars on their shells revealing past traumas. Some scars resulted from cracks made by a predatory crab's attempt to break open the shell. Other scars could be seen as filled-in holes that had been drilled by the toothy radulae of predatory snails.

Vermeij realized that it was easy to see the difference between prey animals that had survived attacks and those that had not. Vermeij also found that by studying shelled animals from different periods of time throughout the past 550 million years, he could see different survival rates. Some periods showed many animals with healed shells, while other periods showed very few healed injuries, hinting that these were periods where survival against predatory attacks was lower.

Vermeij observed that the battles taking place were not yielding consistent kill rates for predators and the same mortality rates for prey. It was from these observations that he proposed a critically important idea: that evolution was like an arms race where organisms are constantly evolving in a bid to outdo their respective predators or prey.

Humans at war have constantly built bigger and deadlier weapons to defeat one another. Hundreds of castles with high stone walls were built during the medieval period. Arrows bounced off their fortifications and it was difficult for troops to breach or climb the walls. Castles were remarkably effective for centuries as defensive structures, and wars often included long sieges while they were starved into surrender because they could not be stormed without losing a lot of soldiers.

Once the cannon was invented, attacks on castles proved devastating – they could be conquered quickly and with little loss of life on the attacking side. The castle became defensively obsolete.

To combat cannons, armies developed rifles that could accurately hit cannon troops far away. They also found that cavalry could race up to cannon groups and kill the soldiers before they could fire. Responding to these new countermeasures, cannons were eventually placed inside protective vehicles that could move around, and the heavy weapons could be aimed in all directions. These became known as tanks. Dr Vermeij found a parallel to this chain of events in the natural world.

Back in the ancient seas during periods when successful predation attempts were common, Vermeij argued that predators must have evolved new characteristics to break the defences of their prey. These innovations were the reason for their success.

Predators and their prey, he argued, were constantly escalating their own arms race, with certain periods dominated by specific adaptations. As soon as predatory snails or predatory crabs developed a new trait that helped them to break shells and feed more effectively, their prey responded by evolving via natural selection. Any prey animals with variations in their armour that helped them to survive the new predatory attack would breed more often and therefore become more common. The prey population would effectively become filled with individuals carrying a more protective shell, forcing predators to adapt to the new defences or find food elsewhere.

The explosion in Cambrian biodiversity appears to have been no different from any other arms race in the way it developed. With the sudden appearance of larger predators like *Anomalocaris*, prey animals were faced with extreme natural selection. Any mechanisms that allowed them to stay alive long enough to reproduce were favoured by natural selection and no group of animals shows this better than the trilobites.

Trilobites are incredibly common in the fossil record: by the end of the Cambrian period and the start of the Ordovician period, 488 million years ago, there were 63 different trilobite families living on the planet. The diversity of trilobites that this represents

Examining fossils with Richard Fortey in a museum in Erfoud, near Mount Issamour.

is mind-boggling. And palaeontologists accept that the fossil record shows only a glimpse of the actual diversity that must have existed among these successful creatures.

Professor Richard Fortey has spent years studying trilobite diversity. This is both an easy and a difficult task – easy because so many trilobite fossils have been found, but difficult because critical parts of trilobites are almost never preserved because they were made of soft tissue.

Fortey adopts a three-pronged technique to deal with this lack of hard evidence. When an enigmatic trilobite appears, Fortey first tries to compare it to living arthropods, such as lobsters, crabs, spiders and insects, for similar body structures. If he finds similarity in the living structures, he infers that the fossilized trilobite may have used its own body structure in a similar way.

With this comparison to modern relatives complete, Fortey ignores the fact that the trilobite is an animal and analyses it as a piece of engineering. By looking at joints in the armour, body structure and form, he estimates the range of motion the animal might have had and how the laws of physics might have affected its behaviour.

Finally, he turns to the environment where the fossil was found. What was this area like when the trilobite was alive? He tries to deduce whether the trilobite was living deep in the ocean or in the sun-soaked shallows of a reef.

The three-pronged approach can be misleading and is no guarantee that a trilobite lived in a certain way, but when all three methods point in the same direction, the argument is more convincing.

We can put Fortey's method of analysis to the test on the trilobite genus *Phacops*. Fossils of these animals are often found curled into balls, reminding us of woodlice. Fortey theorizes that perhaps *Phacops* trilobites were behaving similarly, wrapping their body armour around themselves to shield their soft body parts.

From an engineering perspective, studies of the armour of *Phacops* would suggest that the animal could also have had a flattened form, rather than always being rolled up. Researchers have now collected flat specimens of *Phacops* to confirm the animal could be flat or curled.

Two of the three analysis steps suggest that these trilobites curled defensively when predators attacked. But this is as far as we get because palaeontologists have not yet deduced from the fossil beds where exactly this creature lived. Without more information, we can only speculate what caused *Phacops* to hide its head in its tail and roll into an armoured ball.

———

The trilobites we have met so far have been limited to a life spent crawling the ocean floor, weighed down with heavy, yet necessary, defensive armour. But it was a matter of time before some trilobites,

# Trilobites in Morocco

**"** What images come to mind when you think of Morocco? Busy markets, beaches, camels? It's very unlikely you'd think of trilobites, yet the dusty limestone cliffs of the Atlas Mountains, which form the spine of this country, are the unlikely source of some of the finest trilobite fossils on Earth.

I visited a fossil museum in Erfoud, near Mount Issamour in Morocco, with Professor Richard Fortey, who harbours a life-long passion for trilobites. He discovered his first trilobite fossil at the age of 14, and devoted his career to expanding our knowledge of these fantastic creatures.

Many of the trilobite fossils in the Moroccan cliffs are found curled up, sometimes very tightly with their tails beneath their heads, in a defensive position. Others have their backs arched upwards or contorted into strange positions. It's as though some catastrophe overwhelmed them moments before death.

Professor Fortey thinks he knows the nature of this catastrophe. He believes the seabed, where these trilobites lived, was very steep and made of mud. Over time, sediment would gradually accumulate, building up to a critical point at which a slight movement would cause it to slip and form an underwater avalanche. The mud would have tumbled downhill, carrying with it any inhabiting animals and burying them alive. It sounds rather ghoulish, but it's thanks to these creatures' dramatic deaths and burial in thick, anoxic mud that they are so beautifully preserved, enabling us to study them in detail.

By studying these fossils, experts like Professor Fortey have greatly expanded our knowledge of the secret world of trilobites. New species are being discovered all the time and, at the moment, palaeontologists have defined around 5,000 genera containing many thousands of species. The trilobites were some of the most successful animals in the history of life, and are animals that never cease to fascinate."

**The fossil museum in Erfoud, near Mount Issamour.**

such as the species *Carolinites genacinaca*, started to swim about and try a little hunting, perhaps preying on small plankton.

*Carolinites genacinaca* was alive between 488 and 433 million years ago during the geological period known as the Ordovician. Unlike its crawling kin, *Carolinites genacinaca* had large, bulbous eyes that stuck out in front of its face. The eyes resemble those found on the faces of flying insects, such as dragonflies, and Fortey suspects this indicates a similarity in function.

For animals that swim, being able to look up as well as down is extremely important, to avoid solid objects like rocks jutting up from below and to find food in a three-dimensional environment.

When animals need to move swiftly and engage in actions that require careful coordination, such as hunting, having eyes that face forward with a bit of overlap in the field of vision is very important. Overlapping vision, called binocular vision, allows an animal to perceive distance between objects. This eye arrangement is exactly what humans have.

The forward-facing eyes of *Carolinites genacinaca* suggest that it was a swimmer highly dependent upon vision for its lifestyle. *Carolinites genacinaca* also looks like light infantry, with minimal protection and joints between its armoured plates, which suggests that its body was highly flexible. Mechanically, it looks as if it might have been able to undulate its body and swim highly effectively.

The distribution of *Carolinites genacinaca* in the fossil record also supports the notion: finds in Siberia, Australia, China, the Arctic and the United States could only have occurred if it had been highly mobile.

With light and well-hinged armour, huge eyes and an impressive international distribution, *Carolinites genacinaca* was probably an agile, swimming arthropod that may have spent much of its time pursuing animals smaller than itself at all depths. As with the bottom-dwelling *Cornuproetus*, the dinner menu for predatory *Carolinites genacinaca* is unknown, but it could have fed on some plankton species – plankton known to exist at this time include graptolites – and small arthropods.

——

The prey for other predatory trilobites, however, is less of a mystery. Fossilized tracks found in ancient sediments show that those belonging to the *Olenoides* group often encountered worms burrowing just beneath the ground. The worm-burrowing activity and the trilobite tracks are preserved, so it is possible to see that the two animals were moving along and creating trails. When such trails intersect, the worm-burrowing trails come to a mysterious end and those of the *Olenoides* trilobites continue unaffected. Direct evidence that these trilobites actually ate worms is not present, but it is possible to imagine a sudden cloud of sediment, the thrashing of the worm's tail and then stillness.

Palaeontologists have also studied the fossilized hard parts that sat around the mouths of *Olenoides*; these, too, hint at what *Olenoides* liked for lunch. The gnathobase is a hardened base of an arthropod limb, and in trilobites it can sometimes be preserved. In many trilobites, the gnathobase is relatively smooth and unremarkable, but not so in *Olenoides*. *Olenoides* had dozens of tiny barbs on their gnathobases, suggesting they used them to rummage through muck in search of worms. Since worms are soft-bodied, these barbs would have easily snagged them from the sediment to the trilobite's mouth for feeding.

——

The rocks in Morocco are loaded with valuable fossils, and once discovered, these are often sold on to collectors. When Peter Van Roy, a palaeontologist at Yale University, was a graduate student, he met one such local collector, Mohammed Ben Said Ben Moula. Dr Van Roy deeply valued Ben Moula's ability to find fossils and, over time, they developed a partnership. Ben Moula immediately caught Van Roy's attention when he revealed some impression fossils of soft-bodied animals he had found in local rocks.

The rocks were unquestionably Ordovician, sitting between the period's key dates of 488 and 443 million years ago, yet such soft-bodied animals had never been seen before in such rocks. As the palaeontologists got to work, they identified more than 50 different soft-bodied animals. All of the species were new to science, though their general body forms were not. It quickly became obvious that they were looking at the descendants of the animals prised by Walcott from the Burgess Shale. These animals had not died off at the end of the Cambrian period; they had survived, thrived and even evolved. The only conclusion was that these animals had made it into the Ordovician period without any trouble but, because of their soft bodies, had gone largely undetected by the fossil record.

The new Moroccan fossils came from 40 different localities in southern Morocco, affording the palaeontologists a much wider understanding of life in Ordovician oceans.

With more than 1,500 specimens to analyse, Van Roy, Derek Briggs and their colleagues have their work cut out. It will be years before they can study all of these and report their findings, but even without formally collected data, one thing is obvious: palaeontologists who argued that the Burgess Shale animals were a failed experiment in evolutionary history are now eating their words. The new finds show that underestimating the biases in the fossil record can badly skew interpretations of the history of life.

——

As trilobites developed their own predatory habits, the rest of the animal world moved on. Small and unassuming, our own ancient ancestors were beginning to make their mark. As the Cambrian

A trace fossil held at the Early Life Institute, Northwest University, Xi'an, China, showing a trilobite's last movement across the ocean floor.

Explosion of animal diversity took place, an interesting worm-like animal appeared. Like so many other creatures of its day, it had a hardened part to its anatomy, but its spine marks it as being different. Indeed, it was the very earliest beginnings of the spine that supports the skeleton of so many vertebrate species today.

Charles Walcott first found this odd little animal in 1911, thought it was a worm and formally named it *Pikaia*. In 1979, when Simon Conway Morris took a fresh look at all of Walcott's Burgess Shale fossils, he noticed that *Pikaia* bore a striking resemblance to animals that are alive today. These animals were most definitely not worms.

Conway Morris thought that *Pikaia* closely resembled two families of animal: urochordates and cephalochordates. Urochordates, often called sea squirts, are unremarkable as adults; they are filter feeders and sit on the sea floor, collecting nutrients from moving water. However, as juveniles they are more interesting. Born with little heads and tails and a tiny nerve cord running down their backs, these young animals swim around by pulling on muscles attached to their nerve cord. As they pull these muscles, the contractions cause their tails to swish back and forth, creating propulsion. They use propulsion to find a place to live and grow into adults. As they transform into adults and take on their sedentary lifestyle, they lose their muscles, tail and nerve cord.

Cephalochordates are more complex. They spend both their juvenile and adult lives swimming through the water. Instead of settling down and functioning as filter feeders when adult, they spend their lives searching for small nutrient particles. Like juvenile urochordates, they have nerve cords with attached muscles that allow them to wiggle their tails and propel themselves. However, they also have tiny brains that help them to coordinate their activities by sending electrical signals down the nerve cord.

Few people dispute that animals such as the urochordates and cephalochordates are relatives of our ancient ancestors. The presence of a simple nerve cord, which is a defining characteristic in all fish, amphibians, reptiles, birds and mammals, is evidence enough that these animals are related to the creatures that once formed the base of the family tree to which all back-boned animals (also known as chordates) belong. Even so, for decades palaeontologists resigned themselves to the impossibility of finding our elusive ancestor, since creatures like the urochordates and cephalochordates are soft-bodied, small and extremely unlikely to form fossils. However, the nay-sayers underestimated the high-quality preservations of the Burgess Shale fossil beds.

Conway Morris realized that what he was seeing in *Pikaia* was a rudimentary nerve cord, very similar to the cord found in urochordates and cephalochordates. Based upon comparisons with what is presumed to be its modern kin, he theorized that it may have used its proto-nerve cord for locomotion.

To date, more than a dozen *Pikaia* specimens have been found in Cambrian rocks, and all have shown similar nerve cord characteristics. Walcott's precious, yet overlooked, find was not some sort of aberration.

Some researchers argue that *Pikaia* must have been living worm-like lives in the sediment. Others say that they were using their nerve cords for swimming around and their tiny brains for reacting to predators. It is possible that they sought safety by hiding among larger plant eaters, like trilobites, as the fossil evidence suggests. We will never know their exact behaviours, but whatever they were, they worked. *Pikaia* survived, diversified and, as their descendants made their way into the Ordovician period, evolved dramatically.

The species that scientists believe first evolved directly from *Pikaia* became the first 'fish' in the Earth's oceans. Palaeontologists call them fish but they were not like the fish we know today. For reasons that no one really understands, early fish did not develop backbones to cover the fragile nerve cords inherited from their ancestors. Taking an extraordinary evolutionary digression, it would seem, early fish developed coverings of hard materials composed of phosphate on their heads.

We do not know why the earliest fish developed hard heads and soft backs, but it was obviously a successful strategy. The early fish that arose during the Ordovician period, like the *Astrapis*, ultimately diversified widely through the ancient oceans.

During the Silurian period, which followed the Ordovician and ran from 443 to 416 million years ago, the descendants of *Astrapis* diversified rapidly. Some, like the species *Hemicyclaspis*, became bottom feeders and sucked up nutrients from mud on the ocean floor. Others, like *Anglaspis*, were more streamlined in shape, suggesting that they were speedy swimmers.

Some fish in the Silurian and later periods were simply bizarre. *Doryaspis*, which means 'dart shield', was a fairly short and flattened armoured fish. It had a long, rigid and spiked horn structure protruding from the front of its head shield. It also had hard, stubby and spiked delta-wing-like structures on either side of its body near its plated tail.

Fish belonging to the *Doryaspis* genus were probably bottom feeders because they had mouths facing downward like *Hemicyclaspis* and *Anglaspis*, but the odd, spiked structures on their bodies certainly raise questions. Such a fierce-looking spike on a non-predatory fish may seem rather enigmatic at first, but the feature is not completely unknown to biologists. There's an animal alive today with remarkably similar-looking traits: the sawfish.

# Failed Experiment?

As the arms race went into full throttle and the diversity of animal life exploded with the arrival of hard body parts like those of the trilobites, it has long been assumed that the soft-bodied animals found in the Burgess Shale in Canada became extinct. A few specimens of bizarre soft animals similar to *Hallucigenia* have cropped up in places other than Canada (fossil beds in China have shown that these animals were abundant for the entire Cambrian period), but there has been little fossil evidence to suggest that these animals made it into the Ordovician period.

The explanation that many palaeontologists have traditionally given for the apparent absence of Burgess Shale animals in later rocks is that they were a failed evolutionary experiment. They suggest that natural selection led to the evolution of some body forms in the Cambrian period that worked well in the short term, but failed in the long run. It seems as if the end of the Cambrian period was also the end of most Burgess Shale species.

For decades, this argument has been debated. It has to be remembered that the Burgess Shale fossil beds were extraordinary because they were composed of exceedingly fine sediment, which perfectly preserved the soft bodies of the organisms that died in the ancient ecosystem there. Such beds are extremely rare to find in the fossil record and, for this reason, some palaeontologists have argued that the absence of Burgess Shale soft-bodied animals in rocks from the Ordovician period does not mean they became extinct. Instead, they suggest that Burgess Shale-type animals were present, but they are invisible to palaeontologists today because their bodies were not easily preserved. The Ordovician fossil record is simply biased against soft-bodied animals.

In May 2010, evidence surfaced that challenged the arguments of the failed experiment viewpoint. Peter Van Roy and Derek Briggs, both at Yale University, and a team of colleagues reported in the journal *Nature* that they had discovered more than 1,500 new fossils in Morocco. Many of these ancient animals were clearly related to the animals that Walcott discovered in the Burgess Shale back in the early 20th century.

Unlike the discoveries made by Walcott, these important specimens had already been unearthed and collected by a local Moroccan fossil collector.

Sawfish, or carpenter sharks as they are also known, are members of the skate and ray family, and they have an impressive saw-like protrusion on the front of their faces. This structure is rigid, long and lined with sharp spikes, like the long spear found on *Doryaspis*.

Biologists have discovered that the saws are littered with many cells that can sense electrical activity in the surrounding water. This frightening growth is more like a sonar than a spear because it helps the sawfish to detect animals hidden in the sediment.

The prey may think that by remaining motionless in the mud they are hidden, but they cannot conceal their thumping hearts. Hearts generate rhythmic electrical pulses that the sawfish can detect, thereby revealing their prey's location. The sawfish then dig out the prey with their saws, shovelling sediment up with the flat side of the saw.

But, as we have seen before, animal appendages frequently have more than one function. Unsurprisingly, the saw sometimes becomes a weapon. Biologists scuba diving in the home waters of sawfish have seen them use their saws against swimming prey as well as predators. With prey such as smaller, soft-bodied fish, they swing their saws back and forth, cutting them. With predators such as sharks, the slashing of the saw is a deterrent – why would a shark approach an animal thrashing its metre- (3-ft-) long blade?

It is tantalizing to think of the strange-looking *Doryaspis* as a prototype sawfish, but if it did slash at predators and prey, we will probably never know because the injuries to these soft-bodied animals would be unlikely to be preserved in the fossil record.

So was the protrusion ever used as a spade? Certainly digging behaviours can be preserved in the fossil record as trace fossils, so it is possible that one day palaeontologists may find *Doryaspis* next to some unmistakable excavations.

Most intriguing are the depressions covered with tiny plates of bone in the brain cases of fossil fish closely related to *Doryaspis*. Palaeontologists believe these may be specialized body organs for electrical detection. While these are just theories based upon similarities to sensory organs in modern animals that can detect electricity, such as hammerhead sharks, this is an exciting proposition. If close relatives of *Doryaspis* had electrical sensitivity, it would suggest *Doryaspis* may have had such sensitivities, too.

The looks and behaviour of *Doryaspis* make it appear to be so closely related to the sawfish that it would not be surprising to find it swimming about in our seas today. But, in reality, *Doryaspis*, like *Hemicyclaspis* and *Anglaspis*, was totally different from any fish alive today. Most noticeably, it did not have a moving mouth, just a hole in the front of its head. All food collection was achieved by sucking up water like a vacuum cleaner, which worked well for small food items eaten whole or for small, soft-bodied animals that could be

**Studying the behaviour of the mantis shrimp can give us insights into the life of *Anamalocaris*. It has the essential characteristics of any predator: superb vision, great speed and superior size.**

# Trilobites and Eurypterids

Today, all the trilobites have gone, whittled away by a series of extinction events in the Devonian period and finished off by a mass extinction, which devastated most of the life on Earth around 250 million years ago.

Though the trilobites only remain in the form of fossils today, many arthropod relatives lived on after them. Some of their more ancient cousins grew to a greater size than even the largest of the trilobites. These were the eurypterids, or sea scorpions.

Whilst working on *First Life*, I was lucky enough to film one of these scorpion fossils, which is carefully stored in the vaults of the National Museum of Scotland, in Edinburgh. The fossil is more than 1 m (3.3 ft) in length, and it's quite clear from looking at it that this scorpion was a monster, a terror of the seas. Like the scorpions we know today, it had powerful claws called chelicerae, which it used to grasp its prey: early fish, trilobites and even other sea scorpions.

Complete fossils of these underwater beasts are pretty scarce, but other more cryptic fossils hint at eurypterids much larger than this one. Dr Martin Whyte, a geologist from Sheffield, recently came across a huge sea scorpion track on a Scottish beach, complete with a sweeping trail of the arthropod's tail, fossilized in 330-million-year-old sandstone. Palaeontologists estimate that the scorpion responsible for this trail, known as *Hibbertopterus*, grew to around 1.6 m (5.25 ft) in length and was able to survive for periods out of water.

In terms of size, however, even this monster was outshone by the giant sea scorpion *Jaekelopterus rhenaniae*. Only a single fossilized claw of this aquatic beast has ever been discovered, but it suggests a monster up to 2.5 m (8.2 ft) in length.

With their protective exoskeletons and ability to exploit new ecological niches, it's no surprise that the arthropods conquered the oceans. They continued to thrive unchallenged for hundreds of millions of years, diversifying into the many forms we see today. Arthropods were the true evolutionary pioneers, and would in time conquer a whole new environment: the land.

*Pterygotus anlicus* **was one of the largest of the eurypterids. This specimen, stored in the vaults of the National Museum of Scotland, is a rare example of a complete specimen.**

# The First Sharks

In 2003, a group of palaeontologists, led by Randall Miller at the New Brunswick Museum, reported in the journal *Nature* that they had found the oldest-known shark. At 409 million years in age, the shark was from the early days of the Devonian and more than 14 million years older than any shark discovered to date. This was undoubtedly a shark: it had four rows of sharp teeth clearly resembling the batteries of teeth found in modern sharks, but this creature was not large and fearsome like the great white shark. Known as *Doliodus problematicus*, this shark was the size of a large trout.

Fossils revealing entire shark bodies are very rare because their skeletons are made of cartilage instead of bone. Cartilage decomposes over time and does not fossilize well. Hence, full-body fossils are usually found only in rare soft-sediment beds where impression fossils can form.

Unlike shark skeletons, shark teeth fossilize well and are common in the fossil record. Teeth can help palaeontologists determine what ancient sharks were eating, but only fossils of bodies tell us what they looked like and how they functioned.

Miller's discovery of the full-bodied *Doliodus problematicus* shows that early sharks were already sharp-toothed predators, but small in size. Miller and his colleagues speculate that they probably looked something like the modern angel sharks.

Remarkably, the first half of the *Doliodus problematicus* fossil was found in 1997 by one of Miller's students on a fossil expedition in Canada. Then, two months later, Richard Cloutier, a palaeontologist at the University of Quebec, unknowingly found the second half of the fossil while searching nearby. Much later, Miller and Susan Turner, a shark expert at the Queensland Museum in Australia, realized that not only were the two fossils from the same ancient species, but they were from the same exact animal – truly remarkable.

In many ways, Placodermi, a class of armoured fish, were like the helmet-headed fish of the Ordovician and Silurian seas. They had heavy and rigid head and torso armour but, unlike their predecessors, they had a hinge that allowed them to open and close their armoured mouths. Intriguingly, they also relied on their armour for biting. Instead of having long and sharp teeth, arthrodires, the major group of placoderms that dominated the Devonian period, had sharp edges on the armour next to their mouths. So it was their armour that would have crushed prey animals and splintered their defences.

While many of the placoderms' predators were less than a metre (3 ft) in length, some of these evolved into giants measuring 2.5 m (8 ft) long. The species *Tityosteus* was among the largest and, with sharpened armour on the

**Fish from the Devonian have formidable teeth for attacking prey.**

outside of its mouth, its bite would have proven lethal to all but the best-defended animals.

And so we return to the arms race, around 415 million years ago. The appearance of predators as impressive as *Tityosteus* seems to have put evolution into fast forward, and it is in the seas of the Devonian period that some of the most elaborate defences appear in trilobites.

We have discussed many of the trilobites' defensive tricks: heavy armour, mineral eyes and an ability to curl up into protective balls. But with *Tityosteus* crunching its way through easy trilobite pickings, sharp spines and prickly armour were suddenly in vogue.

The species *Dicranurus monstrosus*, though small, has especially striking spines. Numerous sharp spines stuck out from its body at odd angles, not short and barb-like, but long and wavy. Some of these spines were longer than the entire length of a trilobite's body and, undoubtedly, got in the way of any predator's attempts to bite.

Richard Fortey speculates that the diversification of trilobites with spines was a response to the rise of the predatory arthrodires. With their powerful jaws and rigid armour-based bites, arthrodires would have been able to make short work of most trilobites. It is possible that these predators were specialists at hunting trilobites or other heavily armoured animals in the ancient oceans.

The arms race continued throughout the Devonian. Arthrodires grew steadily larger over time, with the largest species, *Dunkleosteus*, reaching 6 m (19.5 ft) in length. It was similar in size to the largest modern great white shark and bigger than many whale species today. With its vast size and huge, sharp and armoured mouth, it would have made short work of even the most heavily defended animals.

Was *Dunkleosteus* a direct result of an arms race with armoured and spined animals of the Devonian period? We cannot be certain until fossils are uncovered of armoured animals, like trilobites, with bite marks that match those of arthrodires. But size matters – it is one of the simplest ways for predators to bypass powerful defences in prey, but also a tactic used by prey to put off predators.

For their part in the evolutionary story, trilobites themselves eventually gave rise to predatory species. These predatory trilobites also drove the arms race, but in a less noticeable way.

Early predatory fish had hard armour on their heads, leaving a good fossil record. However, trilobites were hunting mainly worms and soft-bodied organisms. These organisms probably evolved in response to the selective pressures exerted by the trilobites. But any changes that did take place are difficult to see in the fossil record because the preservation of these types of animal was so rare. Palaeontologists eagerly await the discovery of the next fossil bed like the Burgess Shale.

slashed by the *Doryaspis* saw. However, these tactics would have been useless on armoured trilobites.

———

We know that the trilobites were unquestionably successful, flourishing for more than 150 million years following the early Cambrian period and appearing in the fossil record over a span of nearly 300 million years. Still, armoured animals would not remain unthreatened for long. During the Devonian period, which came after the Silurian and ran from 416 to 359 million years ago, numerous fish species appeared with powerful jaws that could bite and crush armour with ease. With so many armoured animals in the Silurian waters, there must have been tremendous evolutionary pressure for fish to develop a mechanism to tap into this rich food source. Sucking mechanisms were never going to work well.

Like so many other evolutionary jumps, nobody is yet certain how jaws actually evolved. The fossil record shows only fish that have jaws and those that do not. Fossilized fish with structures that are semi-jaw-like have never been found, although this does not mean that they never will be. Even so, without any transition fossils available, palaeontologists can only speculate at what might have happened.

The current 'best guess' is that jaws actually come from tiny strips of bone called gill arches, which are found inside fish gills. Gills allow fish to collect oxygen from water, and they have associated muscles that evolved to draw water in and out of their mouths and run it past their gills. Palaeontologists speculate that the pumping action of gill arches was taken to an extreme in some Silurian fish that had extremely flexible gill arches that could bite and grasp prey as well.

Until palaeontologists dig up an ancient fish showing a transition stage between being jawless and jawed, it will be difficult to be sure that this gill arch theory is correct. However, the fossil record clearly shows that by the middle of the Devonian period there were some predatory fish with jaws that were being used to cause substantial damage.

When you think of predatory fish, sharks usually come to mind – thanks to films like *Jaws*. But the truly impressive predators of the time were the placoderms, possibly the earliest vertebrates with jaws.

———

But all the evidence of predator and prey interactions found in the Cambrian, Ordovician, Silurian and Devonian rocks pales in comparison with what palaeontologist Simon Braddy at the University of Bristol and a team of colleagues found in a Silurian fossil bed near Prüm, Germany, in 2007.

The team recovered a single claw that ran a scary 46 cm (18 in) in length. After comparing the claw to other species of that day and age, Dr Braddy concluded that the claw had to belong to the arthropod species *Jaekelopterus rhenaniae*, the sea scorpion.

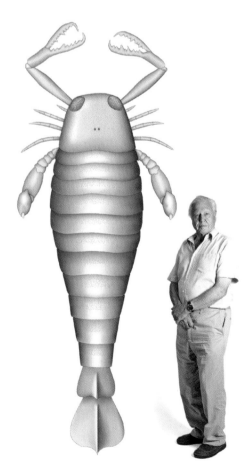

**External skeletons allowed sea scorpions to grow to unprecedented sizes.**

Sea scorpions from the Silurian period were not a new discovery. They were animals with long, large bodies, heavy armour and huge claws. They are thought to have been fearsome predators that ripped their prey into bite-size pieces for their comparatively small mouths. In spite of their name, most of the fossils of these animals have turned up in fossil beds from fresh-water locations like lakes and rivers.

From this Goliath's single claw, Braddy's team could determine the overall size of the sea scorpion by looking at full-bodied fossils of animals of the same species. So the research team pulled out callipers and set to work measuring the claws of other sea scorpions and their total body lengths. They worked out the average size ratio that existed between claw size and body length.

—

*It may seem contradictory that such wondrous diversity is the result of 540 million years of violent hunter and hunted interactions, but the competition between animals seems to be the biggest driver for the amazing natural world.*

Quite conveniently, the ratios of these measurements proved to be nearly constant. Plug in the numbers and what do you get? A *Jaekelopterus rhenaniae* that was potentially 359 cm (11.75 ft) in length – much longer than an adult human. This was an astonishing size, especially for an animal living in a lake, and significantly longer than the next largest sea scorpion ever measured, at 210 cm (7 ft).

Why would these animals grow so large? Their thick armour would have been all but impenetrable; only the largest and strongest-jawed fish would have had a chance – if they could avoid the enormous claws – to get close enough for a bite. However, it is more likely that *Jaekelopterus rhenaniae* grew to this size so that it could eat the large fish swimming around it in the Silurian waters. The sea scorpion was probably the top predator of its day – a shining example of the escalating effect of the evolutionary arms race that began in the Cambrian era.

Though detached from the arms race of the natural world, we can still admire what this constant battle can produce. There are more than 1,000 species of barnacle. It may seem contradictory that such wondrous diversity is the result of 540 million years of violent hunter and hunted interactions, but the competition between animals seems to be the biggest driver for the amazing natural world.

Other theories for how and why explosions in diversity take place will emerge. Some of these theories may demand serious attention and receive the approval of the palaeontological community. However, for now, it seems the incredible diversity of life on Earth is driven by a serious arms race that continually escalates.

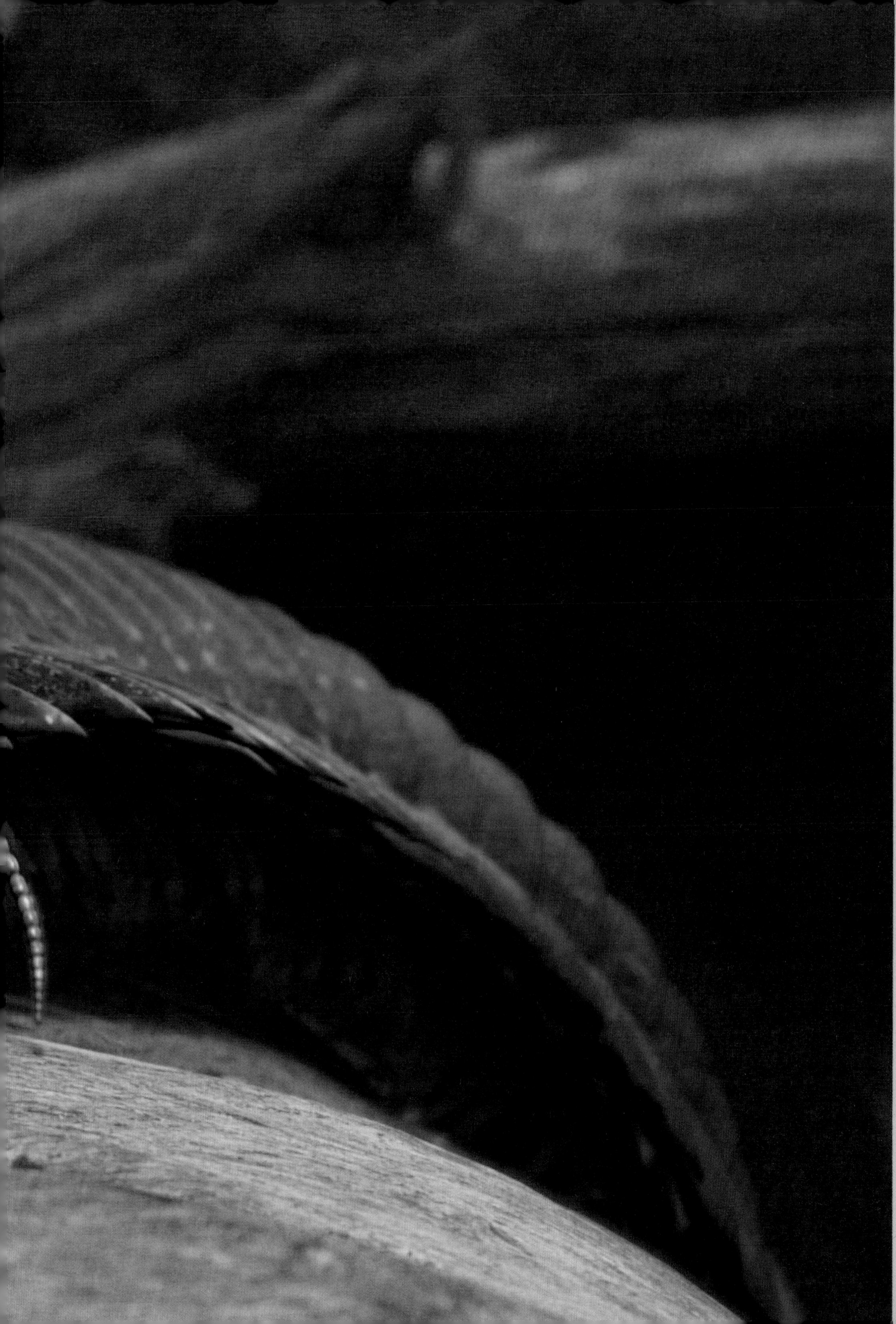

DEEP IN THE tropical forests of Australia, Central America, Asia and Africa, there is a small, inconsequential creature whose body presents a mystery to biologists. Known as *Peripatus*, the animal seems to be a worm, but closer examination reveals a strikingly unworm-like characteristic – more than a dozen stumpy legs.

Some biologists suggest that it is related to the arthropod family, to which the many-legged millipedes, centipedes and insects belong. But arthropods are universally covered with a hard exoskeleton, and *Peripatus* has a soft skin. The biologists who argue that it is most closely related to a worm run into the argument that it has legs, and, of course, worms do not.

To complicate matters further, the genus *Peripatus* contains numerous species, some of which lay eggs with shells around them, while others give live births like mammals. While most worms burrow for food, *Peripatus* is a surface-dwelling night hunter: it crawls around the wet forest floor and squirts a sticky liquid at small arthropods. The liquid quickly solidifies to lock its arthropod prey in place. Once paralysed, *Peripatus* bites through the prey's hard armour

*Peripatus*: the so-called velvet worm is thought to be related to *Aysheaia*, one of the first creatures to move onto the land.

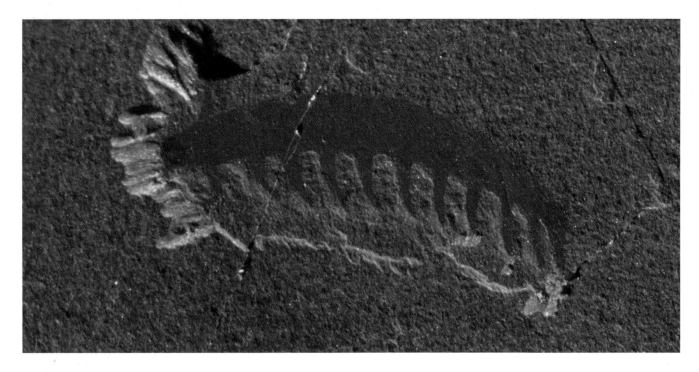

and sucks out its nutrient-rich organs and fluids. Odd as *Peripatus* might seem, it is not simply a modern anomaly. The fossil record suggests that it has been around since the early days of animal life.

In 1911, Charles Doolittle Walcott found an animal with many legs in the 505-million-year-old Burgess Shale. Unsure of what it was, he named it *Aysheaia* after a mountain nearby and tentatively labelled it as a worm. In the decades that followed, arguments arose over whether it was a worm or an arthropod. In the course of time, more fossils of *Aysheaia* turned up. But even with this extra information, *Aysheaia's* affinities to other modern groups remained a matter of debate. Ultimately, George Evelyn Hutchinson, an expert on freshwater biology and ecology at Yale University, examined *Aysheaia* and realized that it looked very like members of the *Peripatus* genus. He proposed that *Peripatus* and *Aysheaia* were members of a unique group of animals unrelated to any other creatures of the modern world.

As similar as *Aysheaia* and *Peripatus* seem, they have one significant difference: the Burgess Shale fossil formation that Walcott discovered was a marine ecosystem, which meant that *Aysheaia* was an ocean-dweller, while the *Peripatus* genus, alive today and well studied, is exclusively land-dwelling. Considering the similarities and close evolutionary relationship, palaeontologists are faced with the question of how and why an animal that was thriving in the oceans of the Cambrian period would ever come ashore.

———

Life in the Cambrian and Ordovician oceans is believed to have shown great diversity as a result of an evolutionary arms race between predators and prey. The fossil record shows that trilobites were filling every available niche and, presumably, other organisms with soft bodies (which would not have been preserved as fossils) were doing the same. Every spot that was advantageous for capturing prey would most likely have had a predator present. We can see strong evidence of this in the ancient fossilized environments, which contain numerous fossilized armoured heads of early predatory fish.

Any section of sediment that contained nutrients would also have had burrowers enjoying the food source. All shallow waters with sunlight trickling down would have been rich in water-dwelling plants. Many assumptions must be made to reconstruct these ancient environments, and while we cannot be sure what the ecosystems looked like, it is easy to imagine them teeming with life.

Yet even with deep oceans and shallow coasts inhabited by swarms of living things, the Earth was still littered with regions that were virtually biological deserts. With only a few hardy bacteria as their residents, the continents that humanity and much of the animal kingdom inhabit today were effectively devoid of animal, plant and fungal life.

For animals, leaving the oceans and exploring the land might seem to alleviate the pressures of crowded and exploited water, but dry land posed new challenges. With bodies that were used to being supported by the buoyancy of water, any Cambrian animal that crawled out of the oceans would have collapsed under its own weight.

Getting enough oxygen while on land was another problem. Animals on land today have specialized lungs that allow them to collect oxygen, whereas Cambrian animals had no ability to collect oxygen from dry air.

There was also the issue of food: animals cannot produce food on their own. They need to eat either plants or other animals, and with no plants or animals present on land, there was no food. So while overcrowding of niches in the oceans was taking place, and while this created pressure to push animals towards the land, the challenges of survival on land were too great and the benefits too small. Conditions on land had to improve before animals made the transition.

———

Plants are a different story. Like animals, they were probably experiencing crowded environments in the Cambrian oceans, and fierce competition, with plants fighting for access to sunlight. In dense tropical rainforests, any time a single tree falls, hundreds of saplings race to fill the sunny spot left behind. Failure to be the first to the top means that a sapling's growth and reproductive success will be jeopardized, which could lead to starvation and death. Thousands of other plant species also fight for light: some attempt to climb existing trees to reach it, some use trees for support, and some attack and kill plants around them to get sunlight for themselves.

In the oceans today, similar fights for sunlight are common. Green algae and seaweed can grow only at very specific depths of water that sunlight can penetrate. Indeed, the intensity of the sunlight is reduced once it enters the water because much of the light is reflected off the water's surface. For this reason, any territory that a plant can claim in shallow water is extremely valuable.

The fact that sunlight travels better through air than water would have created a powerful selective pressure for plants to put some photosynthesizing cells at the water's surface. These cells would have produced far more energy than submerged photosynthesizing cells. Additionally, carbon dioxide, the gas that all plants need to respire, is more readily collected from air than water, and so this would have been another evolutionary incentive for plants to move. However, any plants beginning to explore the water's surface would have faced significant challenges, as their cells would have been threatened by dry air.

Modern plants have numerous traits that prevent them from losing water when they are exposed to dry conditions. They open small, pore-like structures called stomata on the underside of their leaves

# Peripatus

66 Whilst making *First Life*, I got the chance to film an enchanting little creature I have wanted to see for a long time. It is a creature rarely seen by humans, but it is one of great importance. Its name is *Peripatus*, though some call it the velvet worm because its cuticle is incredibly soft to touch.

*Peripatus* lives in locations scattered throughout the tropics, in moist, sheltered habitats, such as rotten logs. This preference for moisture gives just a hint of this creature's ancient past, because it did not evolve on dry land.

If there is such a thing as a living fossil, this surely must be one, because it seems near identical in form to the fossils of the creature *Aysheaia*, found in the Burgess Shale. At first sight, it looks like a worm, but a worm with legs? It has antennae, like an arthropod, but lacks an exoskeleton. The exact definition of *Peripatus* is something of a mystery.

*Aysheaia* lived in the sea, using its tiny hooked feet to cling to ancient sponges as it fed upon them. The modern-day *Peripatus* lives on land, and it has one further attribute that *Aysheaia* could not have had. Tiny holes line its flanks, enabling it to breathe air. These little holes, called spiracles, were first developed by the velvet worms, and they represent a system that all air-breathing arthropods still use today. Air passively diffuses through the spiracles into a network of tubes that extends throughout the animal's body.

It's feasible that velvet worms were the first animals to set foot on land some 540 million years ago, but fossilized evidence of this has yet to be found. Only one thing is certain, once these creatures arrived on land, they hardly changed over the next 500 million years.

Unlike true arthropods, these animals didn't have an exoskeleton. Without a hard exterior they were neither able to increase their size, nor protect themselves from drying out. As a result of this, they have remained tiny creatures and confined themselves to damp environments.

Arthropods, on the other hand, such as the scorpions, a relative of the eurypterids, had impermeable exoskeletons that not only prevented their bodies from drying out but also gave them strength, which they needed if they were to move around without the support of water."

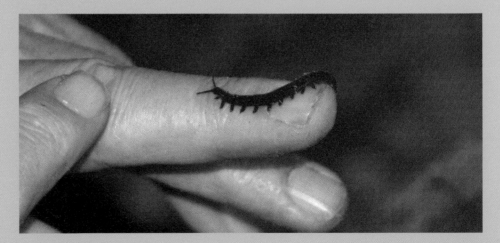

that allow carbon dioxide and oxygen to exchange. If the stomata were to stay open permanently, water vapour would readily exit the plant along with the exhaled oxygen on hot, dry days and leave plants dangerously dehydrated.

Modern plants also have a waxy cuticle on the leaf surface that prevents water from exiting their tissues through evaporation. Plants that live in deserts, where water loss is a problem, have extreme adaptations to their habitat such as thick cuticles and closed stomata during the hottest and driest hours.

Furthermore, as plant cells evolved to take advantage of the sunlight available outside the water, they would have been physically separated from their watery home, so they needed some form of water-transport system.

Early aquatic plants had no need to transfer water through their tissues because all the water and dissolved minerals they needed were readily available in the water around them. Plants on land today have an elaborate root system that draws water from the soil to their photosynthesizing leaves. The system is passive – it requires no input

*Cooksonia paranensis*, an early Devonian flora, was an early land plant with a well developed vascular system for transporting water from the ground to its tissues, allowing it to flourish on land.

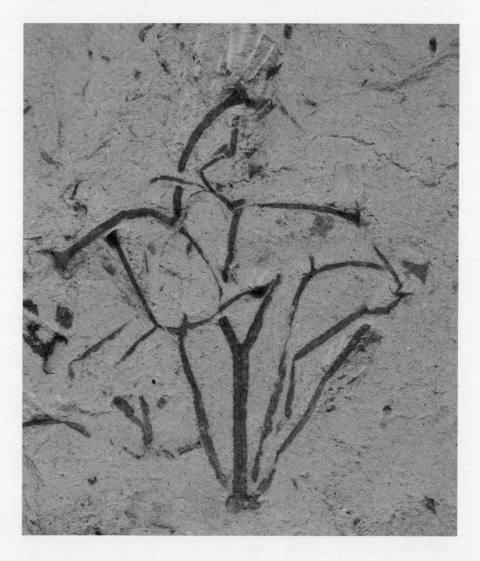

of energy – and thin, hollow tubes in the plants' centre enable lots of water to travel. As water is used and lost through evaporation at the top of the plant, more water is passively drawn up from the roots. It is a perfect partnership, with the specialist photosynthesizing cells in the leaves sharing the energy they generate with the root cells, and they, in turn, share the nutrients and water that they collect from the ground with the leaf cells via hollow tubes.

Early water-dwelling plants would not have had leaves, roots or water-transport mechanisms like modern plants, but they were small enough not to need them. Mosses that are alive today have some specialist tissues that collect water, and other tissues that collect energy from the sun. Instead of hollow tubes for water transport, the cells of mosses are aligned in a grid-like mat. If one cell is dehydrated, then water from a hydrated neighbouring cell moves in to rebalance the overall hydration. With such a system, photosynthesizing cells at the top can be hydrated just by being near saturated cells at the bottom. The result of this is that mosses do not have many cells in between. However, this mechanism is inefficient – it requires mosses to live in wet places and restricts their height – but it is a simple solution to water transport and could have been the evolutionary technique that allowed early plants to pull some photosynthesizing cells from the water and into the air.

Over time, plant cells in the water probably depended upon their surface-dwelling cells for sugars made through photosynthesis, and surface cells probably depended upon the water-dwelling cells for water and nutrients. Such circumstances would have most likely driven these cells to specialize: cells that collected water would become the earliest root cells, and cells that photosynthesized would become leaf cells. As these specializations developed, plants could have migrated from the crowded aquatic environment altogether.

This story of how plants made the journey onto land is highly theoretical because fossils of early plants are rare and the data is limited. Nothing survives from the Cambrian period; the first fossilized land plants are from the Silurian period. The best-known, fully formed terrestrial plant fossil from this time is the modest-looking *Cooksonia*. While *Cooksonia* had its photosynthesizing cells on thin branches rather than on leaves, it was evolutionarily advanced in that it had a rudimentary root system and thin, hollow tubes that look similar to those used by modern plants for water transport – essential for survival on dry land.

Unfortunately, *Cooksonia* reveals little of the evolutionary transition from water to land. The long, thin tube inside its fossilized tissues is already fully developed. There does not appear to be any plant fossil showing the transition stage that must have existed between plants with inefficient water-conducting tissues, and plants like *Cooksonia*.

Even though *Cooksonia* and all other plants from this age lacked leaves – they would not evolve for another 40 million years – the fact that such a well-developed land plant was present during the Silurian period hints that simpler land plants were around much earlier but have yet to be found in the record.

With no plant fossils available from the older Cambrian or Ordovician periods, palaeobotanists have worked with the next best thing: fossilized pollen. Most plants today release eggs and sperm into the air in tiny particles called pollen. Fossilized pollen from the Ordovician period has been known about for decades, and since land plants are the primary producers of pollen, this hinted that land plants existed at that time. Indeed, many palaeobotanists have rallied around pollen findings in Ordovician rocks and argued that these fossils are evidence that early plants were coming ashore. Unfortunately, water-dwelling algae can also produce pollen, and some palaeobotanists suggest that algae were the pollen producers.

Charles Wellman of the University of Sheffield reported a remarkable discovery in 2003 while he and his team searched for pollen fossils in Oman. They were using a series of sieves developed to trap fossilized pollen spores of different sizes. Having found many well-preserved spores, they stumbled upon a series of objects that were elongated and disc-shaped. Dr Wellman and his colleagues realized that they were groups of spores similar to those found throughout the Ordovician period, jammed together inside a plant tissue that resembled the waterproof cuticle of a modern land plant. This fossil finding proved that Ordovician pollen was coming from land plants that were simply not fossilizing very well. More importantly, it provided a glimpse of what early land plants looked like.

As Wellman and his colleagues examined the structure of the walls of the spores, they saw striking similarities between the fossil spores and simple modern plants called liverworts. Like mosses, liverworts move water around their tissues via simple diffusion between adjacent cells. Although a straightforward process, it is inefficient. As a result, liverworts grow only in damp locations, but they are still terrestrial plants.

The Sheffield team's finding in Oman will not end the debate. Sceptics of Ordovician plant evolution will continue to doubt it until more convincing fossilized land plants are found. And with so much pollen turning up in the Ordovician period, more complete plant fossils are sure to be discovered.

———

While the Silurian *Cooksonia* species are thought to have had effective water-distribution systems, they were still primitive and probably struggled to survive on land. Without leaves, they were not maximizing the surface area that they exposed to sunlight. They also did not have stomata or cuticles that could open and close, and they were most likely suffering a great deal of unnecessary water loss.

During the Devonian period, desiccation-resistant traits, such as the formation of cuticles and stomata, started to appear in plant species. These traits helped certain plants survive further from water-drenched locations. This would have been driven by natural selection because, as plants started to expand into dry environments, they could have only survived in really wet soils where water loss from evaporation was compensated by easily available moisture at the roots. These wet soils were prime real estate for early plants and would have become crowded. With increasing crowding would have come the first competitions for access to soil and sunlight on land. Under such circumstances, any plant that could withstand even slightly more arid conditions than its kin could have survived further away from wet soils and endured less competition for resources. This decreased competition would have led to an increased ability to thrive and reproduce, thus giving the plants a higher chance of passing their genes to the next generation.

To outcompete the plants around them, some plants started to grow taller. While Silurian *Cooksonia* specimens were only a centimetre or two (0.5–1 in) in height, many Devonian plants were nearly a metre (3 ft) tall and had leaves on their stems. Gaining height and having leaves would have given competing plants the ability to collect sunlight more effectively and, simultaneously, shade their neighbours. Shade is a weapon that plants use to reduce the amount of sunlight their nearby competitors can collect. However, the benefits of height came with a significant challenge: as plants grew taller, they also had a greater chance of falling over.

To compensate for the problems associated with greater height, some plants, known as progymnosperms, evolved a hard layer of woody tissue around their central stems. These plants, which were the ancestors of pines, redwoods, junipers, spruces and many other modern species, used the woody layer to increase their structural integrity and grow taller without the risk of toppling. With increased height, they also evolved extensive root systems that could grip the ground and function as anchor points.

The woody progymnosperms further improved their ability to grow and compete by developing leaves. Light was originally absorbed by photosynthesizing cells covering the stems but, by the middle of the Devonian period, some plants started developing fern-like leaves that were rich in photosynthesizing cells. This was a phenomenal evolutionary progression, as it allowed plants to increase dramatically the surface areas they exposed to the sun without having to build extra stems from the ground.

With leaves, wooden bark, strong roots and heights as great as 10 m (33 ft), the progymnosperms were effectively the first trees on the

# Forests of Fungi

When Alice met the caterpillar in Wonderland, she was wading through a forest of unusually large mushrooms. The fungi were the size of trees. When Lewis Carroll wrote *Alice in Wonderland* in 1865, the idea of fungi ever growing to such incredible sizes was dismissed as preposterous because modern mushrooms never grow so big. But, in 2000, Charles Wellman, in collaboration with Jane Gray, a palaeobotanist at the University of Oregon with an expertise in early land plants, proposed in the scientific journal *Philosophical Transactions of the Royal Society* that forests of giant fungi might have once been a reality.

Wellman and Gray theorized that a group of plant-like organisms from the Silurian and Devonian periods, collectively known as nematophytes, might have been fungi. Their theory was based upon the fact that the fossils of these organisms, which often look like bundles of fibrous tubes and are unlike any plants alive today, share striking similarities with the structures seen in modern fungi.

One nematophyte, an organism known as *Prototaxites*, had a structure that looked like a tree trunk and was massive. Its trunk size measured nearly 1 m (3 ft) in width and 8 m (26 ft) in length. For decades, researchers have struggled with the fossils of *Prototaxites* and its relatives. Some have suggested that these organisms were just unusual large plants, others that the fossils were the remains of mats of algae, crushed together into a single mass through an obscure fossilization process.

Nematophytes, as a group, have always given palaeontologists headaches because they are a bin of sorts for plant-like fossils from the Silurian and Devonian periods that palaeobotanists cannot identify. New plant-like fossils from these periods are often labelled as 'nematophytes' simply because they do not fit into named groups of plants and fungi. Some may be found to be fungi, some may be plants, and some may be an entirely different group of organisms that no longer exists today.

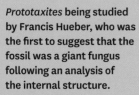

*Prototaxites* **being studied by Francis Hueber, who was the first to suggest that the fossil was a giant fungus following an analysis of the internal structure.**

planet, and as groves of them increased in size, they became the first forests. Other plants that dwelled in these areas were forced to cope with lower levels of light. Those that could survive in shade multiplied and became the first forest ground cover.

Even with strong roots, waxy cuticles and evaporation-regulating stomata, plants during the Devonian period were restricted to damp environments, as they still required considerable amounts of moisture for reproduction. Simple plants like mosses and liverworts release their sexual cells into rainwater and humidity that collects on top of them. These cells then travel through the water to engage in fertilization and create a new plant that, in turn, needs to stay wet during its early stages. As plants migrated from watery environments, this method of reproduction would have become ever harder as conditions became more arid. Ultimately, dry environments would have made it difficult for plants to reproduce and, again, natural selection would have driven evolutionary change. Any plants that could breed and develop as saplings without constant rain and humidity could take advantage of land where no other plants lived. But the evolutionary innovation allowing the expansion of land plants did not come until the Devonian period, when plants started to produce structures that would eventually become seeds.

Seeds today are hard coverings that almost completely surround an embryonic plant. They have two purposes: first, to carry energy so that the embryonic plant can develop a root system and leaves before it has to find nutrients and collect sunlight on its own, and, second, to provide substantial protection. Because they are hard, seeds often prevent animals from eating the vulnerable plant embryo inside, while providing a nearly waterproof layer that keeps it hydrated.

One of the greatest threats to small plants (rather than to large plants) is drying out. The difference in threat level is due to a relationship between the volume of a plant's tissue and the total area of a plant's surface. Larger plants expose proportionally far fewer water-carrying tissues to the surface, experience less resulting evaporation and have a lower risk of death from desiccation. This is why it is easy to kill sapling plants by forgetting to water them for a few days, whereas adult plants are much more forgiving. Larger plants also have a larger body in which to store water when external sources are in short supply.

It is easy to imagine early plant embryos struggling to survive on the edge of moist habitats and regularly dying during dry spells. Under these conditions, any evolutionary characteristics that could help them increase their chances of survival would have been developed, and it appears the evolution of seeds was one of these.

Yet seeds themselves caused a problem. While protection and a short-term supply of food were major boons, the enclosure created by the seed (which formed around the plant egg cell early on) would have

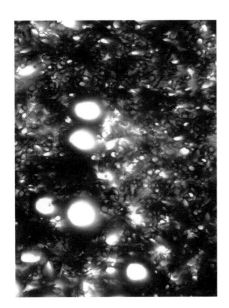

*Prototaxites* is thought to be a very early form of fungus although it has a rudimentary vascular system.

created a barrier to sperm from other nearby plants of the same species, preventing pollination. However, this was not a major issue for the most ancient seeds because they were formed as cup-like structures.

Cup-like seeds sheltered the developing plant embryos, protecting against desiccation, and, simultaneously, provided sperm with easy access to the egg cell. Over time, this protective advantage benefitted the plants that evolved these traits, and seeds became increasingly enclosed. While the protective element evolved, an opening remained at one side of the seed that would still allow sperm to enter the seed and fertilize the egg – a characteristic shown by many modern seeds today.

With seeds, wood, stomata, cuticles, roots, leaves and tissues that allowed effective water distribution, there was no stopping plants from dominating the landscape. While a great deal of evolutionary tinkering would take place, their basic blueprint would remain the same throughout the rest of their biological history.

———

Plants that first invaded the land may have suffered from the harsh and dry conditions, but they also benefited from there being no animals to eat them. However, this quickly changed.

Like plants, animals in the oceans were fighting fiercely for access to resources in an increasingly crowded environment. While plants were competing for sunlight and soil, animals competed for territory and food. Before plant species moved to the surface world, the barren continents could offer animals territory but not food, and simultaneously threatened them with desiccation and asphyxiation. However, once plants made the move to land, both food and territory, which were effectively competition-free, became available on the continents. It is likely this was attractive to plant-eating animals in competitive marine environments, though the presence of food and territory were probably not the only factors.

When plants die, they fall over and bacteria begin to decompose them. This process consumes large amounts of oxygen, and in lakes, bogs and slow-moving rivers, the rotting plant material would have created low-oxygen environments.

The mitochondria inside animal cells require plenty of oxygen. For lake- and pond-dwelling animals during the Devonian period, when plants were growing and rotting in ever-larger numbers, the challenge of steadily lowering levels of oxygen presented an incentive to survive for short periods on the surface. Thus, the rotting tissues of plants likely added to the evolutionary pressures that pushed some animals onto land.

The transition onto land is clearly seen in the fossilized animals found in rocks from the Devonian period. However, identifying exactly when animals became land-dwellers is difficult because, even today, definitions vary about what constitutes a land-dwelling animal.

The mudskipper is a fish that lives in estuaries and coastal areas of tropical and subtropical regions, but it engages in decidedly non-fish-like behaviour under specific conditions. When the tide goes out and most fish hide in rock pools or retreat with the water, mudskippers stay behind.

Mudskippers have strong muscles around their fins that allow them to hop or skip around the muddy terrain at low tide. They also have a limited ability to breathe air. While fish usually depend on their gills to pull oxygen out of water, mudskippers can breathe with their skin, as long as it remains wet. In addition, they inhale a bubble of oxygenated water before the tide goes out, which they hold in their unusually large gills. As they skip around on land, they can use the air in this captured bubble to provide the extra oxygen that is crucial to survival. When breathing water through gills, fish have to work extremely hard to get the oxygen that they need for their tissues because, relatively speaking, water does not carry much oxygen. In contrast, air carries an abundance of oxygen. So, for a mudskipper, a bubble of air held tight in the gills represents a large amount of oxygen.

Rather bizarrely, mudskippers are less active during high tide when they become vulnerable to fully aquatic predators moving into their territory. In order to cope with this threat, they dig into the mud and hide until the tide goes out and they can take advantage of their semi-terrestrial adaptations.

So is a mudskipper a land-dwelling animal or an aquatic animal that acts like one? What is certain is that fish with powerfully built fins

Mudskippers live in estuaries and coastal areas and show characteristics of both land-dwelling and aquatic animals.

along the bottom of their bodies appeared during the Devonian period and it is likely that they were making short journeys on land.

These fish, which are collectively known as the rhipidistians, were predators. Most of them lived in the oceans, but some were found in rocks from estuaries and freshwater rivers. Their fins were intriguingly aligned along the bottom of their bodies, some reinforced with bones attached to their spinal columns. What exactly these fish were doing with strong fins is difficult to determine, but when considered in conjunction with the behaviours of the modern mudskipper, they present a tantalizing story.

Imagine a rhipidistian in a river chasing a small prey animal that is fleeing into shallow water. For the prey animal, moving into shallow water is a brilliant tactic for avoiding large predators because the bulky predator cannot keep up. Now consider the evolutionary arms race and natural selection. If small prey animals were increasingly using shallow water in order to avoid large predators, any rhipidistian predator that could nullify this defence and reach their food would have had a huge advantage. Such a rhipidistian would eat more, live longer, breed more successfully and pass on the characteristic in its genes that helped it to move through shallower water. The characteristic in question looks to have been a mechanism for pushing the rhipidistian off the ground when chasing prey in shallow water.

*If small prey animals were increasingly using shallow water in order to avoid large predators, any rhipidistian predator that could nullify this defence and reach their food would have had a huge advantage.*

As they followed prey into increasingly shallow water, any rhipidistians with fins that could push water, as well as push off the ground, would have been better hunters. The threat of becoming stranded in shallow water would lessen and, more importantly, if they could give themselves extra propulsion by pushing off the ground with their fins, the chances of catching some prey that may otherwise have escaped would have increased.

Rhipidistians must have also encountered increasingly oxygen-poor conditions from the decomposing vegetation in shallower waters. They clearly found a way to manage this challenge because a group of them from the late Devonian period, best known by the genus *Eusthenopteron*, had nostrils on the front of their heads, similar to those seen on dogs, lizards, frogs and people. Moreover, fossils of these advanced fish reveal that they had pathways, which look like small canals, connecting their nostrils with their mouths.

These animals were undoubtedly collecting oxygen from the air and bringing it into their bodies. Where and how this air was being processed is not entirely understood; the best that palaeontologists can do is speculate about the breathing activities of these animals.

———

Whether or not rhipidistians qualify as land-dwellers, there were most definitely four-legged animals present by the end of the Devonian period. In 2006, Edward Daeschler of the Academy of Natural Sciences in Philadelphia, Neil Shubin of the University of Chicago, Farish Jenkins of Harvard University and a team of colleagues reported an astounding find in the journal *Nature*.

Dr Daeschler and Dr Shubin first went to northern Canada in 1999, looking for fossils of early land-walkers after reading about Devonian rocks being exposed on Ellesmere Island. The location presented a challenge: it's an inhospitable icy desert 600 miles from the North Pole and can only be reached by air. Large parts of the island are covered with glaciers and ice, and the average winter temperature is -28°C (-18°F), so it is only possible to go there for a couple of months each year.

However bleak, this site was worthy of extensive study because the Devonian rocks were obviously formed in river and stream environments, where fish were most likely taking their first steps. While searching through the 375-million-year-old river sediments in 2004, the team discovered the bones of an odd-looking creature. The skull of the animal was nearly 20 cm (8 in) in length but, more importantly, the animal had fins on the front of its body, which could almost be considered limbs. They realized that it had existed in a strange twilight zone between being salamander-like in form and fish-like. Its front limb structures had joints in its bones that clearly showed the beginnings of elbow and wrist formation, characteristics that amphibians such as salamanders and frogs do have but that fish do not. Yet these limb-like structures did not have fingers on their ends. Instead, they had tiny fins where digits would normally be present on a salamander.

To add to the strangeness, the fossils that the team discovered revealed the presence of bony scales along the animal's body. Amphibians have soft, non-scaly skin, while fish tend to have scales that are soft and not easily preserved. In contrast, crocodiles, which would not appear on the landscape until nearly 300 million years after the end of the Devonian period and are not closely related to either fish or amphibians, do have scales that are similarly bony in form.

The animal, which the team named *Tiktaalik roseae*, appeared to be a missing link and a critical piece of evidence for palaeontologists trying to understand how fish evolved into terrestrial creatures.

Chronologically, the discovery fitted perfectly into the evolutionary story. Fish, like rhipidistians, with strong, bone-reinforced fins, were present on Earth 385 million years ago, and early amphibians with obvious digits on their limbs were present 265 million years ago. *Tiktaalik roseae* was found in sediment that sat in between these dates and stood as perfect evidence of evolution caught in 'mid-step'.

As for how *Tiktaalik roseae* lived its life, it does not appear that it was particularly adept at either swimming or walking. Its body was heavy and not suited for quick movement on land or through deep water. Instead, palaeontologists speculate it was a shallow-water specialist that behaved like a crocodile. It might have sat in murky waters just offshore and used its long snout to snatch prey passing by in the water and onshore. However, some palaeontologists theorize that it was only coming ashore to avoid the threats presented by larger, fully aquatic predators.

A palaeontologist sketches a 375-million-year-old *Tiktaalik roseae* skeleton.

But fish were not the first animals to come ashore. The oldest-known land animal did not have a backbone, fingers or scales. And it did not take just one single, impressive step either. Instead, it took hundreds of small steps because it was a tiny millipede.

The fossil of this centimetre- (0.5-in-) long millipede was found in 2004 near Cowie Harbour in Scotland by amateur fossil hunter Mike Newman, who realized it was very special. The legs on the fossil were obvious, which suggested that it was a land-walking creature. However, more important were the tiny holes on the animal's armoured body, which looked like spiracles (breathing holes). These suggested to Newman that the animal must have been breathing and living on land. Newman presented the fossil to researchers at the National Museum of Scotland and Yale University.

Palaeontologists at these institutions reported that, while the fragile fossil was an amazing find, its age was what was truly impressive. With rocks in the area being dated to 428 million years old, Newman's fossil looked to be from the middle of the Silurian period, making it more than 50 million years older than the oldest *Tiktaalik roseae* specimens. Arthropods, it seemed, had made the journey onto land long before fish had done so. The researchers named the new millipede species *Pneumodesmus newmani*.

While nobody knows how millipedes made the journey onto land, one possible explanation would be that, as their populations boomed in shallow waters filled with rotten plants, competition for food increased. Consequently, any millipedes that could withstand short periods onshore to feed on dead plant material that was not being eaten by other millipedes would have had an advantage – they could feed more often, breed more often and give their descendants genes that allowed them to survive a terrestrial life.

# First Stumbles

The date when fish made the move to land and became four-legged animals is constantly fluctuating as new fossils are discovered and debated. One of the latest finds, which was reported in *Nature* in early 2010 by Grzegorz Niedźwiedzki, a fossil-footprint specialist at Warsaw University, and Per Ahlberg, a palaeontologist at Uppsala University in Sweden, is moving the discovery of the first footsteps to far earlier than the date of 375 million years ago established by *Tiktaalik roseae*.

While studying 395-million-year-old stone slabs pulled up from a limestone quarry in Poland, Dr Niedźwiedzki and Dr Ahlberg noticed tracks about 15 cm (6 in) wide, apparently made by an animal that was more than 2 m (6.5 ft) in length. Most intriguingly, some of the tracks looked like they had been made by an animal with toes on its feet.

The researchers proposed that their discovery might require rethinking the traditional view that limbs appeared on fish around 375 million years ago with the rise of species like *Tiktaalik roseae*. Instead, they argued that it might be reasonable to consider the possibility that fish made the move to land a lot earlier, and that the current chronology showing landfall taking place between 385 and 365 million years ago was incorrect.

Many researchers have disagreed with the team's interpretation of their finding, mostly because among palaeontologists not all fossils are created equal. While trace fossils like footprints can provide palaeontologists with information about animals that fossilized bones cannot, for many researchers proof of 395-million-year-old, four-legged, land-walking animals needs to come in the form of skeletal fossils. Debates have been rife that the trace fossil could potentially be something other than footprints created by four-legged animals walking across the land.

One of the strongest arguments against Niedźwiedzki and Ahlberg's finding is that fossils in rocks of the same age at the Polish quarry do not reveal evidence of such large-bodied walking animals. The environment that the rocks in the area record was an intertidal lagoon, and the research team is convinced that their fossil represents evidence that there were four-legged animals walking around. Indeed, the researchers argue that careful analysis of the footprints shows evidence of sediments being displaced by the foot of the animal as it set it down with some forward momentum.

What the trace fossils represent will remain uncertain unless the bones of a 395-million-year-old, four-legged land-walker turn up in the Polish Devonian rocks. While the trace-fossil footprints on their own might not be enough to force palaeontologists to rewrite the evolutionary transition of fish to amphibians, the trace fossils accompanied by a skeleton sporting legs most certainly would.

# Making Extraordinary Discoveries

"One thing I love about science is that many crucial discoveries or advances have been carried out or helped along by ordinary people, people who were either extremely lucky or pursued their passion for a particular subject in their own spare time.

Amateur astronomers have contributed to a great number of celestial discoveries, particularly discoveries of supernovae, which are the explosive deaths of stars. By patiently observing the heavens over long periods of time, many people who stargaze as a hobby have spotted very subtle differences in the brightness and size of stars. More often than not, they notify professional astronomers of their sightings, who train highly powered telescopes on the exploding stars, capturing fantastic images of these dramatic scenes.

Many surveys of species diversity are contributed to by ordinary people. Those who love observing birds in their spare time often compile detailed lists of what birds they see, at what time and in which location. If done correctly, these lists, known as censuses, are very helpful to professional ornithologists. They give them an extensive understanding of the distribution of particular bird species throughout the country, or even internationally. Collecting this information would take a single person a long time, but the combined effort of hundreds or thousands of bird enthusiasts makes it possible.

Occasionally lay people can make discoveries that further our scientific knowledge tremendously. Mike Newman is one of those people. He is a bus driver from Aberdeenshire with an extraordinary passion for Palaeozoic fossils. He spends his spare time looking for fossils in the rocks around his home. Scotland is a fantastic place for fossils of ancient fish and arthropods, animals that existed long before the dinosaurs.

Mike found out that the age of the rocks around Cowie Harbour, just down the coast from Aberdeen, had been re-estimated to a much earlier date, and went down to hunt for fossils. What he discovered was truly exceptional. It's a tiny fossil, only about 1 cm (0.5 in) in length. The imprints on it are difficult to make out, but nonetheless it's an extremely significant finding.

Mike recognized it instantly as a myriapod, a millipede or centipede. Above the imprint of the creature's legs are a line of dots, spiracles that would have allowed it to breathe air. This was a creature that walked on land. Why was this discovery so hugely significant? Because the rocks in which he found the fossil are 428 million years old, making this creature more than 10 million years older than any fossilized land creature ever discovered.

Mike duly reported his discovery to professional palaeontologists, who confirmed what he had suspected: it was evidence of the earliest creature to make the great transition out of the seas. This most significant of fossils was named *Pneumodesmus newmani* in his honour.

It's incredibly inspiring to think that we can all play our part in the advancement of human knowledge in one way or another. Armed with a keen interest, any one of us could make a great discovery. All it takes is supreme dedication, the right information and perhaps a little bit of good fortune."

**Examining the tiny *Pneumodesmus newmani* fossil discovered by Mike Newman under a microscope.**

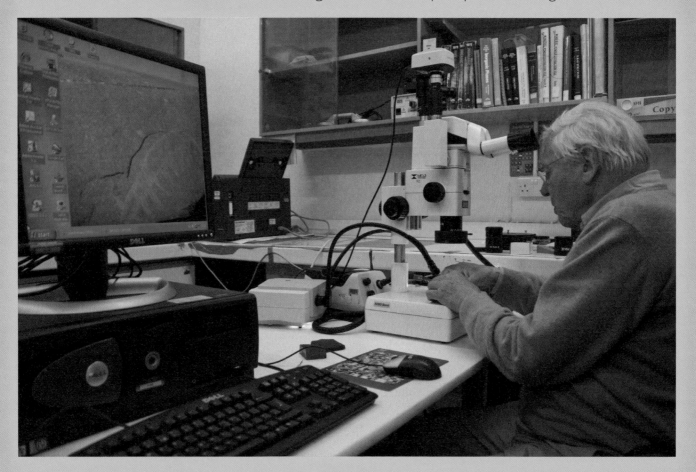

Equally likely is the possibility that predators were the force that drove the Silurian millipedes out of the water. Today, millipedes are hunted by many species, ranging from ants to primates, but the predators that ancient millipedes faced are unknown. What we do know is that the predator threat would have come from the water rather than the land. For this reason, any ability to climb out of the water would have been a major advantage.

The discovery of arthropods climbing out of the water around the time when plants were becoming numerous on land helps explain another reason why fish would have started coming ashore. The rhipidistians were carnivores. If fish with limbs were being lured into shallow water and onto land by food, it was not of the vegetable sort. These animals were predators, and if millipede-like organisms were part of their diet, their presence on dry land would have been another incentive to come ashore.

—

*It is possible that the first arthropod steps on land were taking place not because natural selection was driving animals to find food but because it was causing arthropods to lay their eggs where they would not be predated.*

Yet there is some evidence suggesting that even *Pneumodesmus newmani* may not have been the earliest land-walking animal. Fossil evidence, in the form of tiny rows of footprints, was discovered by Robert MacNaughton of the Geological Survey of Canada in Alberta, Simon Braddy of the University of Bristol and their colleagues. The team was looking for fossils in an inactive quarry 12 miles north of Kingston in Ontario, Canada, when they noticed tracks. Some of the tracks were clear and obvious, seemingly left by animals walking over wet sand. Others were harder to study and are likely to have been made in drier sand that was more readily disturbed by gusts of wind.

All told, the researchers reported in the journal *Geology* in 2002 the discovery of more than 20 trace fossils, and they could see from the tracks that the animals had many legs on their bodies. Some tracks showed evidence of a tail-like structure that must have dragged along behind the animals as they moved over the sand. With body widths of around 8 cm (3 in) and body lengths of about 30 cm (12 in), the team speculated that the animals may have been the ancestors of centipedes, although without fossilized bodies it was hard to be sure.

The tracks fascinated the team because the sandstone rock layer in which they were embedded was believed to be incredibly old.

After extensive study of the region, MacNaugton, Braddy and their colleagues could only prove that the sandstone layer was, at the oldest, 530 million years old, and at the youngest, 475 million years old. Even with such little precision, this proposal was shocking. The date of 530 million years ago was in the middle of the Cambrian period, and the later date of 475 million years ago was during the early part of the Ordovician. With these dates, the finding suggested that animals could have been coming ashore before plants were ever present.

This discovery is still met with scepticism because the specimens are trace fossils, and no fossils of the animals that made them have been found nearby. To complicate matters further, the idea of animals coming ashore before plants were present is difficult to explain. Without plants on land, there would have been no food to lure animals out, and the idea of plants rotting in shallow water and creating anoxic environments that pushed animals out is also unlikely. Predators, however, may have been the selective force that drove animals to make such early landfall.

Today, there are many animal species, including sea turtles, that live at sea and come ashore to reproduce. The evolutionary mechanism of laying eggs onshore probably started in order to protect the young from becoming food for other marine animals.

It is possible that the first arthropod steps on land were taking place not because natural selection was driving animals to find food but because it was causing arthropods to lay their eggs where they would not be predated. At first, the ancestors of the animals that made the tracks that MacNaugton, Braddy and their colleagues found may have been laying their eggs only in shallow water where egg predators were less likely to travel. Then, as the arms race intensified and egg predators moved into these shallow environments, there may have been increased pressure for the egg layers to find ever-shallower locations. Eventually, this may have led to laying eggs on dry land.

The complete story of this process will most likely remain unclear for years, but with so many footprint trace fossils turning up, whatever these animals were doing onshore, they were doing it frequently. More fossils need to be found to support the trackway findings, but if such fossils are discovered, palaeontologists may need to rethink the idea that plants drew animals onto land and be prepared to rewrite much of our early natural history literature.

CHAPTER 11 · TAKING WING: END TO AN ERA

THE ABILITY OF insects to fly is often viewed more with annoyance than wonder as they swarm around or soar over the fences and barriers that we have put up to stop them eating our crops. But despite this common perception, insects' ability to fly is something to marvel at. Aside from birds, bats and pterosaurs (the winged reptiles that lived at the same time as dinosaurs), insects are the only other animal group known to have evolved flight.

This claim often draws cries of protest. What about flying squirrels, flying fish or even flying snakes? While these animals have 'flying' in their name and present the illusion of flight, they are actually gliders.

Flying fish build up speed while swimming underwater and then leap out with their wing-like pectoral fins extended so that they can glide through the air for 40 m (130 ft) or more. Flying squirrels and flying snakes instead make use of the tall trees in which they live. Flying squirrels have membrane-like skin that extends between their limbs and bodies, and creates resistance to the air when they leap off branches. This resistance slows their descent and allows them to glide gracefully from the branches, but they lose altitude rapidly and, unlike true flying animals, cannot gain any lift. Flying snakes use a similar tactic by extending a flap of skin attached to their ribs when they throw themselves from trees.

True flight requires the power of flapping wings, and the fossil record strongly suggests that birds, bats, pterosaurs and insects each evolved wing-powered flight in their own way. In the case of the pterosaur's ancestors, a very long, single finger evolved on each forearm that developed a membrane between it and the animal's body. While these animals are extinct today, they were impressively big when they were alive, with some species growing to the size of small aircraft.

Bats also evolved a membrane, but theirs connected multiple extended digits rather than just one. Birds, however, followed a different evolutionary path. Their entire forearm extended over time and their digits reduced. Instead of having a membrane that filled the space between the extended arm and the body, feathers took up the role of creating a surface area to generate the required resistance. These feathers could then be flapped and used to generate the lift associated with true flight.

*Meganeura*: this giant dragonfly had a wingspan of nearly a metre.

# Giant Insects

" The Carboniferous age was a golden time for arthropods. While some species grew very big, other multi-legged millipedes took advantage of the oxygen-rich atmosphere in a different way. Instead of growing to huge proportions, they remained small, gradually reducing the number of their body segments to increase their agility.

Instead of endless segments and legs they settled on a simple three-segmented body plan: head, thorax and abdomen. The thorax, or middle section, supported three pairs of legs. This body plan may sound familiar – it's one that belongs to a group of animals we encounter on a daily basis. These creatures are some of the most successful creatures on Earth, the insects.

Soon, these tiny insects, only a few millimeters in length, made another dramatic move; they developed wings and became the first animals to fly. The invertebrates had now colonized not only the land but also the air. In this highly oxygenated atmosphere, some of these flying insects became massive.

The most impressive of these flying beasts, *Meganeura*, had a wingspan of nearly 1 m (3.3 ft). That's three times bigger than even the largest of today's flying insects, a beautiful butterfly known as the Queen Alexandra's Birdwing.

This was a marvellous period for the insects, but by no means the peak of their success. Although global cooling and a fall in oxygen levels meant that insects could never retain the huge proportions they reached in the Carboniferous period, these early insects laid down the foundations for what would surely become the most successful group of animals.

Today insects are the most diverse and numerous of all animals on Earth. They have colonized almost every environment, including aquatic environments – both marine and freshwater.

Insects have come up with some of the most extraordinary solutions to life's multiple challenges. Locusts detect when their populations are reaching breaking point, and respond by forming huge swarms in order to exploit new territory. Termites are master engineers, building huge communal colonies that tower above ground and extend even further below it. Ants have mastered the art of farming. Some protect 'herds' of aphids in return for the sweet sap they excrete, and others harvest leaves in order to nourish gardens of edible fungi that they cultivate in underground caverns.

Being relatively large animals ourselves, it is natural for us to think that size is everything. We are wrong. They may be small, but insects are a force to be reckoned with. There are a great many lessons that we have learnt, and still have to learn, from observing them."

While these three evolutionary methods for attaining flight were each incredibly effective, they were a completely different path from the one taken by insects, the first animals to attain flight.

Insects, like all of their arthropod relatives, moult their body's external armour throughout their lives at specific times, in order for them to grow. However, unlike many other arthropods, as insects develop, the forms that they take change dramatically. While young lobsters look much like miniature versions of old lobsters, young insects often have dramatically different physical forms from the adults of their species.

In almost all insects, the juvenile body forms lack wings while adults possess them. These wings are nothing like those of birds, bats or pterosaurs. They are constructed from a wafer-thin layer of rigid but delicate dead tissue that can be flapped rapidly to generate lift. There are numerous insects with juvenile forms that dwell in water and use similar semi-wing structures as paddles to push themselves around. While nobody is really certain whether these paddles ultimately developed into wings, the current thinking is that the paddles found their way into adult insects and started being used for flight.

The benefit of dead tissue making up the wings is that it is exceedingly light, which allows the wings to be flapped very quickly, providing the insect with near-perfect manoeuvrability as a result. The drawback is that wings made of dead tissue cannot be repaired. When bat wing membranes are cut or scratched, they regenerate. When bird feathers break, they grow back. If a wing injury grounds an insect, that insect will never fly again.

From an insect's point of view, however, having such fragile wings is not a problem because insect life strategies are different from those of birds, bats and, presumably, pterosaurs. Birds and bats both grow large in comparison to insects and use vast resources raising their young. Birds build nests and forage for food, which they regularly

*Arthropleura* had a classic multisegmented arthropod body shape. They are likely to have been relatives of modern centipedes and millipedes but grew to around 1.5 metres long.

bring to their youngsters. Bats produce nutritious milk to feed their offspring. Pterosaurs grew to huge sizes, had nests and looked after their young in similar ways to birds and bats. Flying insects, in contrast, do not grow large, do not live for long and do not spend much time looking after their young.

Many insects use their winged adult forms to move around and find suitable mates with which to breed, but most species do not stay alive long after they lay their eggs. In some species, like mayflies, the adult forms do not even have mouthparts or digestive systems. They exist only to breed and lay eggs before dying. Other flying insects use a different strategy. Bees, for example, depend on the formation of a hive in which a non-flying and long-lived queen reproduces, while short-lived flying drones collect food and defend the colony.

———

The largest flying insects that we know of belonged to the genus *Meganeura* and lived in the period following the Devonian. This was known as the Carboniferous period and lasted from about 359 to 299 million years ago. Animals in this group were first discovered in 1880 by the French palaeontologist Charles Jules Edmée Brongniart, an expert on fossil insects at the Natural History Museum in Paris. The fossil found by Edmée Brongniart was enormous for an insect. It had a 75-cm (2.5-ft) wingspan, and mouthparts that clearly revealed it to be a predator. What it was eating is unknown, but it was big enough to be attacking other large insects and, possibly, even some amphibians. Because of the close resemblance that its body had to modern dragonflies, Edmée Brongniart speculated that they might be related. To this day, palaeontologists largely agree that animals in the genus *Meganeura* are giant relatives of the dragonflies that flit with expert precision over lakes and ponds.

The *Meganeura* were not unique among the insects for attaining a large size. Dwelling among the lush forests of the Carboniferous period were multi-legged insects that were most likely relatives of modern centipedes and millipedes but of enormous dimensions. Collectively known as the *Arthropleura*, their size can be deduced from a combination of fossilized footprints and large chunks of fossilized body armour.

With such impressive sizes, palaeontologists have spent a lot of time considering what sorts of lives the *Arthropleura* were living. Were they docile plant-eaters, like modern millipedes, or voracious carnivores, as centipedes are today? Unfortunately, the mouthparts of these animals have not been fossilized, and there is no clear evidence either way. However, in the 1960s, researchers found evidence of plant spores in parts of the fossils where the giant insects' guts would probably have been found. This exciting discovery hinted that the animals were spending at least some of their time nibbling on plants.

# Carboniferous Life

66 Visiting places of palaeontological importance makes you realize just how dramatically our planet has changed over the 4.5 billion years of its history. We filmed in all kinds of locations for *First Life*, some very remote, others a little closer to home. Often it is the less exotic places that resonate and show just how different the world would have appeared millions of years ago.

We visited Crail, a small fishing village on the east coast of Scotland. There is nothing particularly strange about the place until you venture down to the sea. On the rocky shore you can see something that's really extraordinary: a huge circular stump. It looks just like the base of a tree. And indeed that's what it is, or what it was 335 million years ago. This tree wasn't like those we know today. It was related to a group of small, modern-day plants called horsetails, but unlike these small plants it grew to a massive 30 m (98 ft) tall. By any standard, this was immense.

In the Carboniferous period, this area was tropical rainforest. It was a time when the continents of the world were clumped together near the Equator and cloaked in swampy rainforests, nurtured by a humid greenhouse climate.

The abundant plant life pumped out huge quantities of oxygen, changing the entire composition of the atmosphere. Suddenly able to supply their tissues with extra oxygen, the arthropods on land grew dramatically in size and roamed freely through the humid forests.

In the forest that grew near Crail, the ancient trees were rooted in a sandy swamp. The expanses of sand that once stretched between those huge trees have long since turned into sandstone, and it is in these rocks that fossilized tracks have been found. The tracks are in widely spaced pairs, and when you look at them in detail, you can see that each track has a number of dimples in it. These dimples are the imprints of hundreds of individual feet, the feet of a giant millipede-like creature known as *Arthropleura*. Whether this creature fed on plants or animals is not yet known, as a fossilized specimen with jaws has yet to be discovered. What is known, is that it grew up to 2.6 m (8.5 ft) in length, and as such, was the largest terrestrial invertebrate of all time."

*Arthropleura* **thrived in the oxygen-rich atmosphere of the Carboniferous period.**

# Skeletons

" One of the most wonderful aspects of the insect body is their incredibly strong external skeleton, known as the exoskeleton. Just like all arthropods, insects compose their exoskeletons from a tough material called chitin. When creating the exoskeleton, insects have precise control over its composition, and they can inject various proteins and other compounds in order to adjust its flexibility and durability. This ingenious construction provides the insect with protection and, above all, strength.

The rhinoceros beetle is a wonderful example of the strength that the exoskeleton can give to an insect. Rhinoceros beetles get their name from their characteristic and menacing horns, which the males use in intense grappling fights over the mating rites with females. The strength of their horn, which is an extension of the exoskeleton, allows the beetle to flip rival males and can bear incredible weights. But an external skeleton places its own limits on any beetle, in that for all of the protection it offers, it is extremely heavy.

Were a rhinoceros beetle to grow to the size of a rhino, it would barely be able to lift its own legs for the thick and cumbersome skeleton. Another problem imposed by having an exoskeleton is that of shedding. In order for an insect to grow it must first shed its old shell and then grow a new, larger one to cover itself again. After shedding the old skeleton the beetle must wait, motionless, for hours or even days for the new one to harden up.

During this period, the body is not only vulnerable to predation, but is little supported. For a beetle-sized insect, this doesn't pose too much of a problem, but for an insect the size of a rhino, it would not manage to support its own weight, and would collapse.

The rhino, however, has an internal skeleton to give it strength and structure. This skeleton grows with the rhino and is constantly replenished and strengthened throughout the rhino's lifetime. The rhino will never need to shed its skin, and so it is able to grow to an enormous size. At full size, it grows to become the largest terrestrial creature on Earth, a feat that the beetles could never achieve."

**Animals with an internal skeleton, like the rhinoceros, can grow far larger than creatures such as the rhinoceros beetle which has an external skeleton.**

Some palaeontologists argue that the spore evidence is inadequate because the plant spores could have been mixed with the fossils of the *Arthropleura* after the creatures had died. Alternatively, it is possible that, like modern millipedes and centipedes that have relatively similar body forms but very different diets, the *Arthropleura* had considerable diversity, with some species feeding on plants and some feeding on animals.

The existence of large animals like those in the genera *Meganeura* and *Arthropleura* raises an important question. Why were some animals able to grow so large in the Carboniferous period while insects today remain so small? In the decades after Edmée Brongniart made his *Meganeura* discovery, some palaeontologists proposed that insects like *Meganeura* could grow so big because oxygen was in greater supply during the Carboniferous period than it is today. The argument was based on the fact that insects draw oxygen into their bodies through numerous tiny holes in their body armour, called spiracles. This system works for insects because they are not very big, and once oxygen travels through the spiracles, it does not have to go far to get to tissues. The system is passive, and relies on oxygen moving on its own through insect body systems. When oxygen levels in insect tissues fall below the levels in the surrounding air, oxygen naturally moves towards these tissues to bring things into balance. This mechanism of oxygen distribution is different from the active systems that humans have where lungs actively pump oxygen into the body. In the modern environment, insects do not grow larger because their tissues cannot get sufficient and much-needed oxygen into their tissues due to much lower oxygen levels.

During the Carboniferous period, rock chemistry shows that oxygen levels in the atmosphere were much higher than they are today. Oxygen accounts for around 20 per cent of the surrounding air now, but it is estimated to have been between 30 and 35 per cent of the atmosphere in the Carboniferous period, presumably because of the lush rainforests that poured vast quantities of oxygen into the environment. These higher oxygen levels overall may have been what made it possible for oxygen naturally to move deeper into insects' tissues and allowed the bodies of animals like *Meganeura* to grow so big.

The high-oxygen theory has been debated a great deal since *Meganeura* was first discovered. In 1999, Gauthier Chapelle of the Royal Belgian Institute of Natural Sciences in Brussels and Lloyd Peck of the British Antarctic Survey presented evidence in the journal *Nature* that strongly supported the idea that oxygen levels played a role in controlling how large insects became during the Carboniferous period.

Dr Chapelle and Dr Peck were studying small, shrimp-like crustaceans called amphipods that show a tendency to grow larger in cold-water environments than they do in warm. Before they conducted their work, the researchers hypothesized that the bigger individuals in cold waters were the result of the low temperature. During the experiment, the researchers measured temperature and oxygen availability in the water around different populations of amphipods. They discovered that oxygen was the critical factor in controlling body size. As oxygen levels increased in the water, so did the body length of the amphipods. Based on the data from their study, Chapelle and Peck theorized that insects, which are closely related to crustaceans, are likely to be governed by the same restrictions. At the end of their report, they suggested that high atmospheric oxygen levels during the Carboniferous period were probably the reason why giant insects like *Meganeura* and *Arthropleura* existed.

———

*While internal skeletons can literally be thin shafts of hardened materials like bone and cartilage running through the centres of bodies and limbs, an external skeleton must completely surround the tissues that it supports.*

———

While nobody knows for certain what caused insects to lose their ability to grow so large, one theory is that as continents shifted from the tropics, this caused rainforests to became less common. This reduction in the planet's rainforests is suspected to have caused the oxygen content of ambient air to fall by about 15 per cent shortly after the Carboniferous period came to a close. With the drop in oxygen, large insect size probably became unsustainable, as insects could no longer get the vital gas to their tissues. This is thought to have created a selective pressure for insects to become smaller.

However, animals with alternative respiratory and circulatory systems that could inhale air through mouths and nostrils and then rapidly transport gases using blood vessels were unaffected. This situation paved the way for the descendants of shallow-water-dwelling chordates with well-developed respiratory and circulatory systems to take command of the land. Putting respiratory and circulatory systems aside, chordates had another key characteristic that probably helped them to dominate the terrestrial landscape: their backbones.

All arthropods, whether they are crabs, lobsters, shrimps, bumblebees or scorpions, have body armour that makes up an external skeleton. While this exoskeleton plays a key protective role, it is also a critical part of the animals' body structure. Arthropods use their exoskeleton for support in the same way as fish, lizards and humans

# Social Spiders

While colony building might be common in some groups of arthropods, among spiders these traits are nearly unheard of. There are over 39,000 identified spider species in the world, of which just 20 are known to engage in colonial behaviour. Yet when Leticia Avilés, a spider researcher at the University of British Columbia, came across the species *Theridion nigroannulatum* in Ecuador, she could not believe her eyes.

This species of spider lives in a colony of several thousand individuals. Rather than build a large and intricate web to catch insects as they fly by, these tiny spiders hang a series of silk strands down from leaves. Once they have hung their strands, they hide upside down beneath plant leaves and wait for their silk strands to catch unwary prey. As soon as a flying insect is snagged by the hanging strands, the tiny spiders leap down in an impressive ambush. They mob the trapped insect and quickly throw more strands up and over their prey, while simultaneously biting the animal with their venomous fangs.

In many ways, they are like tiny nomadic hunters because they are unable to carry their large kills back to their nests. Instead, they work together to carry their prey back. With particularly large prey, Dr Avilés also saw for the first time the animals taking turns carrying their food back.

An intriguing mystery surrounding these spiders is that their colonies are not always made up of thousands – many colonies are made up of just 20 to 40 individuals. What controls the differences in colony size is still unknown.

The tendency to come together continues in other species. Birds form colonies where individuals help look after each other's nests; fish form shimmering schools that make it much harder for predators to attack; and mammals, like prairie dogs, form extensive towns that help them keep a wary eye out for danger. Remarkably, it is humans that show some of the strongest similarities to the hive behaviours of insects.

Just take a moment to think about it. How many people do you know who go out and harvest their own crops, slaughter their own animals, manage their own waste and build their own homes? Even if you do know one or two completely self-sufficient farmers, it is a safe bet that they regularly use power tools, telephones and ovens during their daily lives and that they probably do not know how to build these devices from scratch. Most people in developed countries turn to supermarkets for food, water treatment services for waste management, and engineering companies to build their homes because the people who run these services are specialized at what they do. Even farmers, who can do a great many things on their own, depend upon outside services for certain things that make their lives simpler. Like hive-dwelling insects, we live in a society where we are highly dependent upon a system of divided labour with different individuals specializing at providing specific services.

**Young Garden Orb spiderlings (*Araneus diadematus*) in France.**

**Communal spiders' webs in the Southern Highlands, Papua New Guinea.**

use their internal skeleton. Just as these vertebrates would collapse if their internal skeletons were removed, arthropods would not be able to support their bodies without their armour. However, as similar as the internal skeletons of chordates and the external armour of arthropods are, there are two key physical differences.

The first major difference is that arthropods must moult their external skeletons in order to grow. As they gain enough nutrients to grow bigger, arthropods shed their old external skeleton, at the same time growing a new, larger skeleton. Although effective, this process has many associated costs. As the arthropods have to make a new exoskeleton from scratch, they must collect sufficient nutrients so that they can grow and discard skeletons throughout their lives. The process becomes more and more expensive because their skeleton must get larger and stronger to protect and support their increased size. Vertebrates, in contrast, do not have these issues. There is no expansion taking place on the inside of their bones so, as growth occurs, the skeleton inside the body thickens and lengthens without having to be discarded entirely and regrown.

—

*A giraffe can grow continuously throughout its life by making comparatively minor adjustments to its skeleton and it never has to shed its thick protective skin. Ladybirds must shed their skins regularly in order to grow and also endure occasional periods of vulnerability.*

The process of making the skeleton grow larger certainly has nutrient implications as well, but it is nowhere near as costly to the animal doing the growing. In addition, moulting also forces arthropods to endure a brief period when their new, larger exoskeleton is soft and pliable, which makes them vulnerable to predation.

The second key difference between chordate internal skeletons and arthropod external armour is associated with the fact that external armour is extremely heavy because of the many surfaces that it must cover. While internal skeletons can literally be thin shafts of hardened materials like bone and cartilage running through the centres of bodies and limbs, an external skeleton must completely surround the tissues that it supports. Just think about it for a moment. If your arm were surrounded by bone instead of being supported by two thin bones on the inside, you would have far more bone present and, more problematic, all of that extra bone would be very heavy.

A simple comparison between a vertebrate with an internal skeleton, like a giraffe, and an insect with an exoskeleton, like a ladybird, shows the price that arthropods pay for their external armour. A giraffe can grow continuously throughout its life by making comparatively minor adjustments to its skeleton and it never has to shed its thick protective skin. Ladybirds must shed their skins regularly in order to grow and also endure occasional periods of vulnerability. Most importantly, if a ladybird could somehow get enough oxygen into its tissues to grow to the size of a giraffe, its armour would be so thick and heavy that it would not be able to do much except look menacing – just imagine a 3-m- (10-ft-) high ladybird. As a result of their more effective respiratory systems and their internal skeletons, the vertebrates took over as the key large land animals as the Carboniferous period came to a close.

———

It may be tempting to think of bigger animals as being better than small ones and to consider arthropods as having failed the trials of evolution. Indeed, it is fair to say that their external armour restricts their evolutionary growth such that they cannot easily evolve and fill the niches that larger animals like elephants and giraffes do. However, even with the restrictions of their skeleton, and respiratory and circulatory systems, from an evolutionary perspective, arthropods are a remarkably successful group of animals.

Having evolved all the way back in the Cambrian period, arthropods as a group have survived every single mass extinction on the planet. Certainly, individual species fell by the wayside at specific points in history, but when the great Permian extinction wiped out 90 per cent of all the families on the planet 251 million years ago, many arthropods endured. When the dinosaurs and many other organisms were wiped out at the end of the Cretaceous period 65 million years ago, the arthropods survived yet again.

Merely surviving mass extinctions is not the sole trait that marks arthropods as extraordinary – sponges, cnidarians and echinoderms all survived these extinctions, too. The key difference between these other groups and arthropods is the number of habitats in which they are able to thrive.

Today, arthropods have colonized nearly all corners of the globe: there are crabs that climb palm trees to collect coconuts, and lobsters that live in arctic waters littered with icebergs. There are also spiders that build trap doors in desert sands to catch unwary prey; spiders that build nets with their silk to collect air bubbles so that they can breathe underwater; and dragonfly larvae that are powerful enough to catch fish. While these animals are never the largest in their ecosystems, they can be found almost anywhere on the planet, from the highest mountains to the densest forests.

Remarkably, even with the physical limitations imposed on them by their exoskeleton, some arthropods have found ways to function as if they were much larger organisms than they actually are. They do this by forming colonies.

Consider a beehive. To an ordinary observer, each individual bee is an individual animal, free to do whatever it pleases. This is, however, not strictly correct. Most bees living in a hive are unable to breed. They exist only to collect pollen and defend the hive. When they attack, they release a venomous stinger that may kill the attacked animal but will definitely kill the individual bee – honeybees cannot live without their stingers. These bees are the sacrificial pawns, defending the colony from danger.

Inside the hive there is a single queen bee that spends her day laying eggs and being tended by workers. By being fed different types of nutrients during development, the larvae that form from the eggs develop into different types of adult bees. The vast majority of these larval bees become non-reproductive workers that collect food and defend the hive; others become males, and a few become queens.

Like pollen being released into the air by plants to fertilize other plants far away, certain males fly off to distant hives where they can share their sperm with other queens. This creates diversity in the honeybee population and is vitally important to the evolutionary

**Thousands of individuals have cooperated to build these termite nests in Australia. By living in vast colonies, termites have become super-organisms, working and behaving as one.**

survival of the species. Sperm transport is effectively the male's only task in life. Similarly, queens born into a hive fly far away and build new hives – if you ever find a small piece of honeycomb in an attic, you can be sure that there was once a young queen trying to establish a hive of her own.

The large number of workers that a single queen is able to produce means that a solitary hive can have a wondrous effect on the surrounding landscape. The bees collect nectar from flowers and, while doing so, they become covered in pollen that they unwittingly transport to other flowers in the area, helping to fertilize them. The relationship is a beneficial one for both plants and bees: the bees receive food in the form of nectar, and the plants are sexually fertilized.

In contrast, the presence of a termite or an ant colony can be extremely destructive, as the colonies voraciously consume resources. A single colony of insects can have a similar impact on an area as a large animal.

A critical question that biologists often ask is whether the individual bees are the animals or whether the hive is the animal. Certainly, all of the bees in the colony have mouths of their own and can metabolize nutrients without outside assistance. But the workers bring food to the colony, and the queens and males do all the reproduction. Without the workers, the hive would have no food supply and be defenceless. But without the queen and males, the hive would cease to reproduce.

So even though they are unable to grow individually to a large size, insects have evolved colonial behaviour that places single animals in the colony into similar roles as individual cells or organs in a body. In a way, the specialization seen among individuals in insect colonies is like the specialization that took place with single-celled organisms that formed the first sponges. Just as the single-celled organisms were able to benefit by sticking together in the ancient oceans following the Snowball Earth events, there must have been some early insects that started working together in quasi-hive-like ways that proved evolutionarily very effective.

———

It is a long journey from the evolution of the fractal fronds in the Mistaken Point fossil beds to the development of the first ecosystems on land but, remarkably, this journey is only the first chapter of several in the evolution of animal life on the planet. The 250 million years that followed the age of giant insects would see giant reptiles swimming the seas, pterosaurs soaring across the skies and dinosaurs roaming the land. At first glance, it might seem that this world of reptiles was completely different from the world that existed earlier, but it was not. If life were a play in a theatre, it would be safe to say that the cast of characters during the dinosaur era was different from the earlier eras, but the plot was much the same.

Contrary to how they are commonly viewed and portrayed in films, dinosaurs did not start out huge. At the beginning of the dinosaur era, the plant-eaters and meat-eaters were all relatively small and similar in size. Then, at the start of the Triassic period, dinosaurs entered an arms race with ever-more elaborate mechanisms of attack and defence. And just as the mild-mannered *Pikaia* proved to be an unsung underdog of the Cambrian period and would eventually lead to the evolution of fish, which would dominate the seas, the dinosaur era had its unsung underdogs in the form of the earliest mammals. These early mammals were almost all unimpressive rodent-like creatures, most likely scurrying beneath the feet of the giant reptiles. Nevertheless, mammals would eventually have their day.

The giant dinosaurs died out 65 million years ago, possibly as the result of some great environmental catastrophe. Only the dinosaur's feathery descendants, the birds, passed into the most recent evolutionary chapter that became dominated by mammals.

Mammals are very different from reptiles. Most give birth to live young instead of laying eggs. They feed their young with milk produced from their bodies. They also maintain a constantly warm body temperature, which makes it possible for them to remain active

in cold environments where most reptiles cannot function – there are no known reptiles able to live in arctic environments. The exception is birds, which technically are reptiles but they have developed constantly warm body temperatures, too. Despite all the differences, mammal evolution played out very similarly to dinosaur evolution and to evolution during the Cambrian, Ordovician and Silurian periods.

Mammals started out small but quickly grew in size. Predatory mammals developed new mechanisms for attacking their prey, and prey mammals developed new mechanisms for defending themselves. While some of the defence mechanisms differed, the majority were similar. Consider the tactic of rolling up into a ball for protection. Trilobites evolved this tactic and so have armadillos.

Such similar solutions to common problems have arisen again and again throughout evolutionary history. The rounded and bulbous eyes of the trilobite *Carolinites genacinaca* helped it to see all around its environment, just as bulbous eyes on flying insects provide them with panoramic vision. The long and slender snout of the early land-walking animal *Tiktaalik roseae* was probably used to snatch prey while sitting in shallow waters, in the same way as crocodiles use their snouts today.

Natural selection will continue to drive the evolution of life on the planet for millennia to come and, based upon what palaeontologists know about the past, many millions of years from now more stories with different casts of characters should continue to be told. Of course, the nature of evolution, and the diversity of life on Earth in the future, will be influenced by the activities of our own species.

While life on Earth has survived meteor impacts, snowball episodes, dramatic shifts in climate and numerous other environmental catastrophes, never before has a single species on the planet held so much sway over so many others. With the invention of the combustion engine, humanity unknowingly created an easy way to make the Earth warm up very quickly, producing a man-made greenhouse effect. With the invention of nuclear weapons, it became possible to create nuclear chain reactions across the planet that could have devastating effects upon living things. With genetic engineering and the building of the first artificial life with human-created DNA, the potential to create new types of animals that could wildly change evolution on the planet is a very real possibility.

The power to control whether the same old story will continue or whether it will change dramatically in the millennia to come rests in human hands. We have the unprecedented ability to determine the future story of life, as well as the responsibility to ensure that it doesn't end with us.

# index

mitochondria 49, 49-50, 51, 92, 192-3, 250
molluscs 159
Moon 25
Morocco 15, 16, 55, 192, 192, 193, 201, 220, 221, 223, 229
Morris, Simon Conway 184, 185, 187, 189, 194, 195, 226-7
mosses 245, 246, 249
Moula, Mohammed Ben Said Ben 225
Mount Issamour, Morocco 16, 192, 221, 223
movement, first animal 128-9, 141, 147, 152-4, 156-7, 159, 179, 180-1, 211
mudskipper 251, 251, 254
multi-cellular life, evolution of 19, 21, 36-7, 43, 47, 81, 92-100, 103-5, 112, 125, 127, 137, 150, 155, 192, 193
mutation, genetic 36-7, 40, 45, 49, 52, 103, 122, 161, 175
mysid shrimp 96-7
—
Namibia 72
Narbonne, Guy 79, 108, 109, 113, 121, 122, 127, 128, 154-5, 155
natural selection 40, 170-1, 174, 175-6, 180, 181, 217, 221, 229, 247, 249, 254, 261
Nedin, Christopher 203, 207, 210, 211
nematocysts 104-5
nematophytes 248
nerve cord, evolution of 226-7
new species, finding 146
New Walk Museum, Leicester 11
Newman, Mike 256, 259
Niedzwiedzki, Grzegorz 257
nucleic acids 32, 51
—
oceans:
  acidification of 91
  animals move out of onto land 241-2, 250-1, 254-61
  arthropods conquer 233
  iron in 47, 85
  life begins in 14, 20-2, 23, 25, 32
  overcrowding in 241-2
  plants move out of onto land 242-50
Olenoides 225
Oman 246
On the Origin of Species 117

Opabinia 168-9, 168-9
ordinary people, extraordinary discoveries by 258-9
Ordovician period 221, 224, 227, 229, 234, 236, 241, 246, 261, 283
organelles 49-50
Ottoia 211, 212-13, 213
Owen, Richard 29
oxygen:
  banded-iron formations and 46, 47, 47, 85
  cyanobacteria and consumption of 88
  in Earth's ancient atmosphere 45-6, 47, 85
  Europa 25
  giant insect growth and levels of atmospheric Carboniferous 274-5
  human lung and 160
  mitochondrial consumption of 92
  movement of plants and animals from ocean to land and 250, 251, 254, 255
  post-ice age rise in levels of 88, 92, 93, 96
  worm absorption of 152-3, 156-7
—
Paine, Robert 216, 217
palaeomagnetics 65, 67
palaeontology, adventures in 82-4, 82, 84
parrots 75, 76, 77, 80, 175, 176-7, 178
Peck, Lloyd 274
penis worm 144, 150-1, 211
Peripatus 240-1, 240, 243, 243
Permian period 176, 177, 178, 279
Peytoia nathorsti 195
Phacops 222
photosynthesis 41, 43, 47, 72, 88, 117, 179, 242, 244-5, 247
Pikaia 218-19, 226-7
'pizza disc' 118-19, 119
Placodermi 234
plankton 47, 104, 224
plants 45, 46, 117
  Cambrian ocean 242
  height of 247
  land, move out of ocean onto 242-50, 244
  progymnosperms 247, 249
  stomata 244, 247, 249
  water-transport system 242, 244-5, 246-7

plate tectonics 17
Pneumodesmus newmani 256, 259, 260
population booms 72-3, 88, 217, 256 see also Cambrian Explosion
Porter, Susannah 81, 83, 85
Precambrian period 7, 9, 38, 113, 119, 142, 174
predators 190-213
  Anomalocaris as first 170, 193, 195, 196-7, 198-9, 201, 203, 205, 206, 207-8, 209, 210, 210-11, 211, 212, 213, 221
  Cambrian Explosion as a result of the rise of the first animal 216-17, 220-37, 241, 254, 282
  emergence of 193, 195, 196-7, 198-9, 203, 205, 207-8, 209, 210-11, 213, 216-17, 220-37, 241, 254, 282
  fuels arms race of diversification 220-37, 241, 254, 282
  increase diversity in an ecosystem 216-17, 220-37, 241, 254, 282
Priapulida 150-1
progymnosperms 247-8
protoplasts 88
Prototaxites 248
pterosaurs 264, 268, 269
—
Queen Alexandra's Birdwing 267
—
radial symmetry 112, 113, 132, 159, 195
radula 159
Ramsköld, Lars 189
red rust 47, 85
reproduction 32-3
  birth of 32-3
  clones 52, 53, 55
  corals 163
  evolution of 15, 52-3, 52, 55
  fossilization of 163
  Funisia and 162, 163
  RNA and 32-3
  speed of specialization in animals and 160-1
  variation and 52-3, 77, 103, 122, 163, 175
respiratory system, evolution of 152-3, 156, 157
rhino 272, 273
rhinoceros beetle 272, 273, 273

rhipidistians 254-5, 256, 260
RNA 32-3, 33
—
Saturn 25
sawfish 227, 232
scientific debate 29
sea pen 8, 117-18, 120
sea scorpion 233, 236, 237
sea squirts 226
seeds 249-50
segmentation 153, 156
sexual reproduction: clones 52, 53, 55
  corals 163
  evolution of 15, 52-3, 52, 55
  fossilization of 163
  Funisia and 162, 163
  speed of specialization in animals and 160-1
  variation and 52-3, 77, 103, 122, 163, 175
Shackleton, Sir Ernest 63
sharks, first 234, 234, 236
Shubin, Neil 255
silica 19
Silurian period 227, 234, 236, 237, 245, 246, 248, 256, 283
single-celled life 14, 19, 20, 21, 40, 43, 46, 150, 161, 198, 251
  begin consuming other single-celled organisms 47, 49, 50, 92, 103-4, 105, 192-3
  evolution into multi-cellular life 96, 99, 161
  fed solely by volcanic nutrients 36
  move to dependence on sunlight 37
  power of 43
  sex and 55
  Snowball Earth and 72-3, 74, 75, 81, 85,
skeleton 226, 236, 272, 272, 273, 273, 274, 275, 278-9
Slushball Earth 68
Snowball Earth 57, 58-85, 59, 60, 61, 62, 63, 64, 66, 67, 68, 70, 71, 73, 283
South Africa 40, 41, 44
speciation 75, 77, 80-1
spermatozoa 104, 125
spiders, social 277
Spitsbergen Islands 67
sponge 92-100, 96-7, 98-9, 104, 104, 105, 112, 113, 141, 148, 153, 193, 195, 279
Sprigg, Reginald 113, 140, 145

HarperCollins*Publishers*
77-85 Fulham Palace Road
London W6 8JB

www.harpercollins.co.uk

Collins is a registered trademark of HarperCollins*Publishers* Ltd.

First published in 2010

Introduction © Sir David Attenborough, 2010

Text, Visuals and Illustrations, with the exception of those detailed below © Atlantic Productions (Chevalier) Ltd, 2010

Thanks to the following for permission to reproduce imagery: p23: Deborah Kelley and Mitch Elend, University of Washington, Seattle. First published in Science and reprinted with permission from AAAS; p24 © Michael Benson/Kinetikon Pictures/Corbis; p33 © Dr Elena Kiseleva/Science Photo Library; p36 © P H Plailly/Eurelios/Science Photo Library; p41 © Dr Jeremy Burgess/Science Photo Library; p42 © Derek Lovley/Kazem Kashefi/Science Photo Library; p43 © Sinclair Stammers/Science Photo Library; p44 © Georgette Douwma/Science Photo Library; p48 © Karen Brzys, Gitche Gumee Museum/www.agatelady.com; p49 © Professors P Motta & T Naguro/Science Photo Library; p51 © Dr David Furness, Keele University/Science Photo Library; p52 © Visuals Unlimited/Corbis; p60 © Dr Juerg Alean/Science Photo Library; p62 © Bernhard Edmaier/Science Photo Library; p63 © Science Photo Library; p66 © Julian Calverley/Corbis; p68 © Joseph Kirschvink; p74 © Eye of Science/Science Photo Library; p80 © Natural History Museum, London; p84 © Tony Craddock/Science Photo Library; p93 © J Gross, Biozentrum/Science Photo Library; pp94-5 © H R Bramaz/PSI; p97 © Jean Vacalet; p101 © Michele Westmorland/Science Faction/Corbis; p102 © Alexis Rosenfeld/Science Photo Library; p104 © Peter Scoones/Science Photo Library; pp123-4 © Philip Donoghue, University of Bristol; p128 © Alex Liu, University of Oxford; pp186, 212 © Alan Sirulnikoff/Science Photo Library; p225: Courtesy of Qiang OU, originally published in Palaios; p230 © Fred McConnaughey/Science Photo Library; p234 © New Brunswick Museum, Saint John, New Brunswick; p236 © Simon Braddy, University of Bristol/Robert Brady; p244 © Dr Philippe Gerrienne, University of Liege. First published in the Review of Palaeobotany and Palynology; p248 © Carol Hotton, National Museum of Natural History, Smithsonian Institution, Washington; p251 © Jürgen & Christine Sohns/FLPA ; p276 © Jean Paul Ferrero/Ardea; p277 © Bob Gibbons/FLPA. Computer-generated imagery on pages 2-3, 9, 12-13, 21, 26-7, 34-5, 64 top and bottom, 86-7, 106-7, 118-9, 120, 130-1, 135, 138-9, 149, 161, 162, 164-5, 168-9, 182-3, 188 top and bottom, 190-1, 194, 196-7, 209, 212 bottom, 214-5, 218-9, 238-9, 252-3, 262-3, 266-7, 268, 270-1 created by ZOO (www.zoovfx.com). Photographs on pages 173 and 282 taken by Gary Moyes.

15 14 13 12 11 10 9 8 7 6 5 4 3 2 1

A catalogue record for this book is available from the British Library.

ISBN 978 0 00 736524 1

Associate Publisher: Myles Archibald
Senior Editor: Julia Koppitz
Senior Project Editor: Helen Hawksfield
Design and layout: Steve Boggs & Terence Caven
Production: Steven White
Colour reproduction by Dot Gradations
Printed and bound in Italy by L.E.G.O.

# acknowledgements

THIS BOOK, AND the television series it accompanies, have been possible due to the extraordinary insights that have been made in recent years into the earliest animal life on Earth. We hope this book has captured how our understanding of early life is continuing to evolve through new fossil discoveries, new analytical techniques and new ways of bringing those early animals back to life through computer-generated imagery.

Such advances have come through the work of palaeontologists across the world, often working in remote locations and in extreme conditions to further their science. We have drawn widely on their published work in books and academic journals. We would like to thank all the scientists who have been directly involved in the book and series. They have been generous in their time and highly supportive of our attempt to bring together their work for a wider audience. We would especially like to thank Hazel Barton, Jean-Bernard Caron, David Deamer, Bernard Degnan, Philip Donoghue, Mary Droser, Richard Fortey, James Gehling, Justin Marshall, Guy Narbonne, Susannah Porter, Mike Russell, Geerat Vermeij and Ken Verosub. We are particularly grateful to Patricia Vickers-Rich for her help throughout the series.

Many of the scientists were kind enough to read and comment on early drafts of the sections of the book relevant to their research. We are especially grateful to Richard Cowen, at University of California, Davis, and Edwin Colyer for reading the entire first draft manuscript and giving invaluable input.

Thanks must go to the team at Atlantic Productions without whom this book would not have been possible, in particular the series producer, Anthony Geffen, and the series director, Martin Williams. James Taylor has been project manager for the book. Josh Young drew on the series to write additional material. Peter Johonnett, associate producer for the series, read and commented on early drafts while Will Benson and Hannah Sneyd provided additional research. Naturally, the book benefited from the work done by the production team creating the series, particularly the contributions made by Candice Martin, Kirsty Wilson, Ruth Sessions, Olwyn Silvester, Peter Miller, David Lee, Tim Walker, Peter Hayns, Paul Williams, Matt Pearson and Mimi Gilligan.

During the filming of the series, the production team were assisted around the world by a large number of individuals and institutions who facilitated access to the latest finds and most important sites. Particular thanks go to Coralie Armstrong, Anna Catchpole, Sandie Degnan, David Harris, Hamish Haslop, Michelle Larsen, Anne Miller, Suzanne Miller, Jan Perry, Don Reid, Dennis Rice, Jovanka Ristich, Belinda Ryan, Chris and Evan Ryan, Ian Smith in Australia; Joe Jazvac, Alex Kolesch, Chris McClean and Richard Thomas in Canada; Brian and Mina Eberhardie, Hammi ait Hssain, Robert Johnson, Brahim Tahiri in Morocco; Marco Stampanoni in Switzerland; and Simon Braddy, Emma Butt, Jon Clatworthy, Hannah Dolby, Mark Evan, Mark Graham, Jack Jarvis, Ruth Robinson, Andrew Rorison, Andrew Ross, Clare Torney, Louise Wilce in the UK. Our thanks also go to Susan Attenborough for her help throughout the production.

The stunning photo-realistic computer-generated imagery of the animals reproduced in the book was created by the specialist visual effects studio ZOO, who collaborated closely with the scientists to ensure that the recreations of the creatures and their environments are as accurate as possible. The team was led by James Prosser and included Rob Angol, Matt Baker-Jones, Carlos Cidrais, Sam Cox, Aleksandra Czenczek, Jay Harwood, Ryan Lewis, Craig Mepham, Phinnaeus O'Connor, Bhaumik Patel, Simon Reid and Rhiannon Tate. Most of the location photographs were taken by Anthony Geffen.

The series was made possible by the support of broadcast partners around the world including Jana Bennett, Alan Yentob, Janice Hadlow, Kim Shillinglaw, Cassian Harrison and Carol Sennett at the BBC; Mark Reynolds and Suzanne McKenna at BBC Worldwide; Jonathan Blyth at 2 Entertain; Clark Bunting, Stephen Reverand, Bill Howard, Susan Winslow and Sarah Hume at Discovery Communications; and Marena Manzoufas at ABC Australia.

For the book, our thanks to Jonathan Lloyd at Curtis Brown. A large team at HarperCollins has worked on the production of the book. We are grateful for the contributions made by our editors Myles Archibald and Helen Hawksfield and designer Steve Boggs, as well as Hannah MacDonald, Iain MacGregor and Julia Koppitz. Personal thanks from Matt Kaplan to Thalia.

Finally, and most importantly, the thanks of everyone involved in *First Life* go to David Attenborough. This project grew out of his passion for the subject and he has travelled thousands of miles around the world to reach the most important and remote sites. Throughout, David has remained enthusiastic and committed and inspired everyone who's worked with him.